Innovations and Techno-ecological Transition

Smart Innovation Set

coordinated by
Dimitri Uzunidis

Volume 7

Innovations and Techno-ecological Transition

Fabienne Picard
Corinne Tanguy

iSTE

WILEY

First published 2016 in Great Britain and the United States by ISTE Ltd and John Wiley & Sons, Inc.

ISTE Ltd
27-37 St George's Road
London SW19 4EU
UK

www.iste.co.uk

John Wiley & Sons, Inc.
111 River Street
Hoboken, NJ 07030
USA

www.wiley.com

Library of Congress Control Number: 2016948015

British Library Cataloguing-in-Publication Data
A CIP record for this book is available from the British Library
ISBN 978-1-84821-876-5

Contents

Preface

In August 2015, the French Parliament adopted the law on the energy transition for green growth after a year of citizen debates. A few months later, Paris hosted the 21st UN Climate Change Conference (COP 21). These two key events provided an opportunity for gaining an individual and collective awareness about the impact of our activities on our environment and of their consequences. They remind us of the limited natural resources that are used extensively in our daily lives. They highlight the shortcomings of post-industrial societies, stressing that solutions exist or are emerging within various communities to build a low carbon society.

As quoted by Albert Einstein, "*The significant problems we have cannot be solved at the same level of thinking with which we created them*". Exiting the current paradigm is the goal of the underlying sustainable transitions discussed here. How can we encourage this structural transformation and the emergence of socio-technical systems that are environmentally friendly? How can we create a dialogue between technological innovation and the environment in order to reconcile man and nature, the economy and the environment?

These questions could not be missing from the collection of works published by ISTE. We propose a systemic vision built around the process of innovation that goes beyond disciplinary boundaries by focusing on two major areas, energy and the agriculture-food industry. In writing this book, we are aware of our privileged position, that of researchers living in developed countries, where access to water and electricity is instantaneous for a relatively low cost. It is in this particular posture that we write without trying to give a universal character to our remarks.

This book is the result of work done within the Research on Innovation Network (http://2ri.eu) whose objective is three-fold: to observe and analyse the process of innovation, theorize innovation systems and value research in economics and management of innovation. We thank Dimitri Uzunidis, its President, for giving us the opportunity to write this book and all of our colleagues whose ideas have stimulated our thinking.

Fabienne PICARD
Corinne TANGUY
August 2016

Introduction

In its initial meaning, the term transition defines the physical change of state of a substance moving, for example, from a liquid to a gaseous state. It was then applied to the analysis of other types of systems: social systems (transition from an agrarian society to a market society), political systems (transition of communist countries to a market economy) and more recently to technological systems. The transitions discussed in this book relate to the field of sustainable development and its three pillars, which are economic, social and environmental. They are considered sustainable transitions or sustainability transition.

The year 2015 was marked in France by the promulgation of the law on energy transition (18 August 2015) and COP21 in Paris. For the first of these, its objective is to modify the impact of human activities on the environment (the stated objectives of the law on the energy transition are to make buildings and economic housing efficient in terms of energy consumption, giving priority to the development of clean transport, achieving zero waste and making today's wastes tomorrow's materials, continuing the growth of renewable energy, fighting against energy precariousness), and for the second, its objective is to limit climate change to a rise in global temperature of 2°C by the end of the century compared to the pre-industrial period.[1] These events question the implementation of structural changes (transition) that will make socio-technical systems respectful of the emerging and environment, and thus create a dialogue of

1 The synthesis of the IPCC (Intergovernmental Panel on Climate Change) Fifth Report published in November 2014 states that if emissions continue at the current pace, rising temperatures will instead be 4. 8°C.

innovations, technology, and environment, and reconcile man and nature, the economy, society and the environment. This is, in other words, to achieve a balance of the three constituent pillars of sustainable development to ensure sustainability of the implemented solutions.

Resource management practices, individual behaviours, organizations and other forms of groups, which structure the contemporary patterns of consumption and production, are tested here. The issue is the transformation of our societies to make them resilient, inclusive and sustainable. The challenge is matched only by the complexity of the subject and the abundance of literature that seeks to understand, analyse and to propose solutions.

When discussing the concept of sustainable development, a definition often put forward is "*A [mode of] development that meets the needs of the present without compromising the ability of future generations to meet their own needs*" according to Mrs. Brundtland, Prime Minister of Norway and President of the World Commission on the Environment and Development, given in 1987 in the report Our Common Future (Brundtland report). What can be considered as "sustainable" is certainly subject to discussion, and the awareness of the need to establish immediately global solidarity, beyond intergenerational solidarity, appears increasingly strong. This process was initiated following a series of alerts. Economic alerts, such as the one initiated in 1971 by The Limits to Growth by the Club of Rome[2], taken in 1972 to the United Nations Conference on the Human Environment in Stockholm where it was then a question of eco-development. Ecological alerts about environmental and climatic imbalances (acid rain, hole in the ozone layer, melting glaciers, deforestation, etc.), the multiplication of industrial disasters, Seveso (1976), Amoco Cadiz (1978) and Chernobyl (1986), Exxon Valdez (1989), and more recently Fukushima (2011). Social alerts related to health risks.

Gradually, instead of the opposition between economy, ecology and growth, a reflection on the relationship between these concepts and terms of sustainable integration of these three dimensions (economic, social and environmental) has taken place. It is clear that "*our growth patterns are not sustainable given the resources and limits of the planet; we must go through*

2 The report on the limits to growth (The Limits To Growth), also known as the Meadows report, is a requested report to a team of Massachusetts Institute of Technology by the Club of Rome in the early 1970.

a transition to rebuild our models and achieve sustainable development"[3]. The dualistic vision opposing economy, growth and ecology, environment, which still appears in the different transition scenarios [DAV 14], leaves room for the expression of other views of society. Systemic approaches to transition emphasize the fact that the structural changes that appear unavoidable today cannot be solely borne by the development of new technologies or the introduction of technologically innovative solutions, nor by the way of the market. Social models that are being redefined will be the result of multidimensional developments where economic, technological, sociological and environmental constraints intersect in a multi-stakeholders? co-construction process.

Beyond the simple greening of the current model of society, the ecological transition is based on two inseparable components: "*social and societal innovation [...]: the ecological transition implies the emergence of a new governance, new ways of acting, producing, new and more sober consumption practices that are built and shared by all the stakeholders to gradually build new collective references; technological innovation and research and development in the fields of organization and industrial processes: it is necessary to work on all the modalities to save natural resources and reduce environmental impacts. This is particularly the case for sectors with a low rate of infrastructure and equipment (production of energy, construction, transport, etc.) renewal, for which the choices of the next few years will be crucial to influence the long term trajectory*"[4]. However, does innovation allow the achievement of environmental objectives cheaper and faster? Can eco-innovations form the basis of a new model of society?

The underlying assumptions for our purpose are the following:

– The ecological transition can be increased by new technologies, but it cannot be reduced to this one technological dimension

– Technologies are primarily social constructs and innovation is by its nature a systemic process

– Energy, mobility based and agro-ecological transitions constitute an inseparable converging whole, carrying the same citizen social remobilization.

---

3 http://www.developpement-durable.gouv.fr/Qu-est-ce-que-la-transition.html.
4 http://www.developpement-durable.gouv.fr/Qu-est-ce-que-la-transition.html.

To address the issue of the relationship between innovations and transitions, this book is divided into three chapters. The first chapter aims to set the scene, both theoretical and factual, of the issue of sustainable transition and its necessity. In doing so, it shows that the structural change of a company can be analysed through technological, societal and institutional innovations. This applies to all the major social functions, that is to say socially structuring economic activities, which fall within "innovation systems". Two of the major functions receive special attention, and these from the following two chapters: first of all the energy, then agriculture and food processing sectors. Not only are these societal functions vital to maintaining the human condition, but they are supported by combinations of territorial scale where tensions between the local and the global are expressed. In doing so, they invite us to question the "classical" models of innovation in the context of socio-technical transitions. We will seize the chance in each of these chapters to review the conditions of implementation of these innovations, but also the blocking and locking factors that hinder the success of the transition. The conclusion will question the territoriality of the studied processes and review the nature of the required changes.

1

A Necessary Transition?

The aim of this chapter is to demonstrate the empirical and theoretical components that influence the reflection on sustainable transitions today, that is to say the structural transformations of socio-technical systems. In a prosaic way, the concept of transition, moving towards sustainable development, can be defined as the structural transformation of society, or its constituting subsystems, towards modes of production, distribution and consumption that are more respectful to the environment and less energy-consuming, notably of fossil fuels and natural resources [OEC 11]. More than a state to be achieved, the transition can be seen as a process of moving to a different society, a process consisting of various routes.

This structural transformation of societies can be "spontaneous". It can also be guided, directed by "survival" imperatives and a collective awareness of the need to preserve the environment and natural resources. It then becomes a subject of debate and potential expression of a citizen draft [SCA 15]. Sustainable transition becomes a political and social project; it takes the quality of an ecological or socio-ecological transition. Hidden behind these different terms is a particular vision of the processes of construction and transformation of societies, a different interpretation of the relationship between the environment, technology, human and society (in its economic, political, socio-cultural).

If sustainable transition is discussed in this work through the light of transformations in socio-technical systems, it also implies that these major structural changes are made in a multi-dimensional dynamic setting, involving technology, society and the environment. The question of how these transformations occur then arises, the respective place occupied by the different dimensions of these systems, the nature and forms of their interaction, the addressed locks and encountered barriers.

Not only do systems have to change in order to meet the different challenges that they face, but the implementation of a transition[1] to a to a low carbon society is associated with an evolving relationship with the environment and nature. In a society that is totally mediated by (technological) objects [KAP 09], the transition urges us to question the place and the evolution of these objects and the arrangements that are implemented in order to address major societal functions. By redefining the human's relationship with technological systems that are environmentally focused and not exclusively anthropocentric, we rethink the definition of the three pillars of sustainable development and their combination.

To answer this question, we quickly review the limits of socio-technical systems and transition issues (section 1.1). We will then see how these new concerns invite researchers, especially in economics and sociology, to propose a new framework of transition analysis and innovation processes (section 1.2) in which eco-innovations and environmental innovations take place (section 1.3).

1.1. Socio-technical systems facing their limits

In socially structuring areas such as energy, transport, and food, the socio-technical systems in place have reached their limits and the macro-environment's increasing pressure makes adjustments at the margin increasingly inadequate. Thus, the energy supply system is faced with a depletion of natural resources, especially fossil fuels, air pollution (local and global) and emissions of greenhouse gases, but also a nuclear risk, revived by the Fukushima accident in Japan in 2011, the difficulties of securing energy and raw material supply in a context of geopolitical instability, energy insecurity of a growing part of the population [INT 11]. Transport must also cope, especially in major cities, with road traffic congestion, increased local air pollution, depletion of fossil fuels, especially hydrocarbons oil, which impacts the fuel price, increased CO_2 and greenhouse gas emissions (Box 1.1), and a growth in the number of accidents [GEE 12]. As for agriculture and the food industry, they must also face many difficulties: loss of biodiversity and repetitive food crises.

1 This expression refers to the need to 'decarbonize' our production methods by reducing the use of fossil fuels (coal, oil or natural gas) whose combustion emits harmful particles, including carbon dioxide (CO_2), one of the greenhouse gases responsible for accelerating climate change.

The reflections on the need to operate transformations in the modes of production and consumption are related to the awareness of the existence of a system that reached its limits. These limits and dysfunctions of the current socio-technical systems appear during intense recurrent crises of varying intensity. They affect the capacity of these systems to meet the large societal functions assigned to them; providing humans with food, housing and transportation. Authors like Grin *et al.* [GRI 10, p. 1] go even further by considering that "*... without such a shift to a more sustainable economy, we might also not be able to solve the financial and economic crisis in the long run*". But more than an absolute limit, what we face is the incompatibility between the mode of development initiated in the wake of the First Industrial Revolution and current socio-technical systems used that appear to be less and less able to provide solutions to the demographic shock of the late 20th Century.

1.1.1. *Meeting global demographic pressures*

Global population pressure probably constitutes one of the first break points. If the population growth rate was relatively contained from the Neolithic period to the First Industrial Revolution, the latter, in only 200 years, increased the world's population from just one billion to more than seven billion people today according to the latest forecast of United Nations [UN 15] and INED [INE 13].

Over the last centuries, and especially in the 20th Century, humans have developed all kinds of socio-technical systems, contributing to the improvement of their living standards and life expectancy, but have ignored environmental limits and available and accessible natural resources[2]. The use of natural resources for productive purposes and satisfaction of human needs makes the human living conditions dependent on their ability to exploit these resources. As Rotillon [ROT 10] recalls in his introductory remarks, the economic study of the problems related to the exploitation of natural resources by humans leads to a dual concern of resource depletion and of environmental degradation.

2 A natural resource exists, from an economic point of view, only if it is useful to man (which means becoming aware of its existence and knowing to use it). Man will use it with a given technology; "We speak of natural resources in the economic sense when the resource will be used with the existing technology and is exploitable with current prices" [ROT 10] thus, fisheries, forests, coal, oil, plant, water ... are natural resources.

It is observed that a large part of these challenges is environmental in nature, even if they encompass strong economic and sociological issues. Indeed, most of the technological solutions used today in key areas of energy and agriculture/food processing are an important source of negative externalities on the environment. The economists define externalities or external effects as a production or consumption activity of an agent that affects the well-being of another without either of these receiving a compensation for this effect. Pollution in all its forms is a typical example of a negative externality: when a factory is emitting waste into the environment, it provokes, without compensation, a nuisance to local residents. Traffic congestion is an example of a reciprocal negative externality: each motorist affects and is affected negatively by the other [HEN 16].

1.1.2. *Limiting the depletion of natural resources*

The growing use of natural resources inevitably leads to depletion when we refer to non-renewable resources. A corollary to this orchestrated scarcity is that resource accessibility is reduced and costs increase. The impact of the physical exhaustion of natural resources on economic growth is immediate (steady state, zero growth, degrowth).

It is interesting to note that the classical economists, in the 19th Century, analysed the consequences of the depletion of natural resources on the economic development. Thus, Ricardo [RIC 17] envisaged a steady economic state under the constraint of the decrease in the fertility of available arable land. Malthus [MAL 20] considered, meanwhile, that population growth was inconsistent with available resources. As he pointed out to Jevons, in 'the coal question' (1865), the end of the Industrial Revolution is nearing in England because of the exhaustion of coal deposits. In 1912 the Italian chemist Giacomo Ciamician recalled that modern civilisation was the result of fossil coal that man has greedily exploited but that these deposits were not inexhaustible [VEN 05]. As Albrecht notes [ALB 09], the appearance of fossil fuels was initially accompanied by an awareness of the shortage and the finite nature of this resource, which did not stop man from building a civilisation based on these resources. This issue was obscured from 1930 only to "reappear in 1970 at the publication of works of the Club of Rome (Meadows Report [MEA 72]). A few years later, the oil shock following the Kippur War (1973) raised the issue of energy independence and the securing of supply, resulting in the adoption of the Messmer Plan for the deployment of nuclear power plants in France (1974–1986). Beyond the availability of natural resources, their destocking generates environmental damage that intensifies the increase of exploitation of these resources, in connection with the growing demand generated by the increasing demographic pressure previously mentioned.

1.1.3. *Restrain environmental degradation*

Environmental degradation caused by anthropocentric patterns of production and consumption covers a broad spectrum. The latter extends from the alterations of local ecosystems (local pollution of extraction sites of natural resources, pollution of groundwater, deforestation) to the risks incurred locally by the people as in the case of shale gas extraction [WOE 15], to more global effects. One of the most emblematic manifestations of anthropogenic effects is undoubtedly global warming. Greenhouse gases (GHGs; Box 1.1) are the link between human activities and global warming. These gases are certainly present in nature, but a growing proportion of greenhouse gases results from human activities. Their accumulation since the beginning of the industrial era (the life of a CO_2 molecule is a century) is a major accelerator of global warming as it intensifies the greenhouse effect. The increase in the greenhouse effect is due to the increasing concentration of GHGs in the atmosphere, resulting in an imbalance of heat exchange between the Earth and space, thus contributing to global warming. It is estimated, since measurements of the global average surface temperature of the Earth were established, that this temperature has increased by about 0.85°C between 1880 and 2012 [IPC 13]. The ocean temperature has also increased and land glaciers have melted.

In this matter, alarm was expressed by the IPCC. Created in 1988 with the initiative of the United Nations, the Intergovernmental Panel on Climate Change (IPCC) was tasked to assess – in an unbiased, methodical and objective way – the scientific, technical and socio-economic available information in connection with the issue of climate change[3]. It proposed a methodology for evaluating emissions of GHGs by country. Its various reports have contributed to the awareness of the anthropogenic nature of climate change and the need to act given the consequences (economic,

3 http://www.developpement-durable.gouv.fr/Presentation-du-GIEC.html.

political and social) of non-action. The irreversibility of the processes and the slow trend of reversal mechanisms suggest significant ecosystem reconfigurations on Earth. Rapidly introducing structural changes became compulsory and that is the aim of transitions hoping to achieve more durability. In June 1992, the Framework Convention on Climate Change in Rio pointed out the need to stabilize the concentration of greenhouse gases in the atmosphere "*at a level that would prevent dangerous interference with the climate system and in a sufficiently rapid manner to allow the adaptation of savings, preservation of food production and the establishment of sustainable economic development*". As far as we know, the international negotiation processes fully illustrate the difficulty of implementing global governance, and the tragedy of the commons developed by Oström (Nobel Prize in Economics in 2009).

The greenhouse phenomenon is a natural one. It involves the heat exchange between Earth and space. In this process, the Earth receives and absorbs energy, primarily due to solar radiation. Part of this radiation is reflected by clouds, Earth's surface and, oceans out into space, and some radiation is absorbed by the atmosphere. Some gases in the atmosphere absorb this thermal radiation and re-emit the heat to the Earth's surface: this is called the greenhouse effect.

The absence of this greenhouse effect would result in an average temperature of −18°C on Earth. The increase in the greenhouse effect leads to a rise in the average temperature of the Earth's surface. A number of gases (carbon dioxide, methane, ozone and artificial gases, fluorinated gases such as chlorofluorocarbons (CFCs), perfluorocarbons (PFCs)) accumulate in the atmospheric layers and increase the greenhouse effect. These gases, known as greenhouse gases, trap thermal infrared radiation emitted from the surface of the Earth and change its "radiation balance", that is to say, the balance between the energy absorbed by the Earth and emitted out to space.

Apart from carbon dioxide (CO_2), which accounts for 70% of GHG emissions originating from anthropogenic sources, mainly from the combustion of fossil fuels and biomass, the IPCC identifies about 40 greenhouse gases. Among the most important is nitrous oxide (N_2O), which constitutes 16% of emissions resulting from agricultural activities and biomass combustion. Methane (CH_4) represents 13% of the emissions. It originates from agriculture, landfills, production activities and energy distribution. Fluorinated gases (HFCs, PFCs, SF6) are, in turn, used in refrigeration systems, aerosols and the last two in the semiconductor industry. Although they account for only 2% of emissions, these gases have a higher per-molecule impact than CO_2. Different GHGs are differentiated by their degree of nuisance and their lifetime. Methane has a global warming potential 25 times greater than CO_2. This impact is measured by a Global Warming Potential Index (GWP) in 100 years: a GWP of 1 for CO_2, 25 for methane, 298 for nitrous oxide, from 7,400–12,200 for perfluorocarbons and 120–14,800 for hydrofluorocarbons. No GWP is given to water vapour, which has the distinction of not staying more than 2 weeks in the atmosphere therefore not contributing to long-term global warming.

Box 1.1. *The greenhouse effect and the role of greenhouse gases*

1.2. An analytical framework under construction: the Transition Studies

The aim of research in the process of sustainable transitions is to study the forms of responses that are made to solve major challenges that modern societies are facing. From a scientific point of view, the research trends that are interested in these questions are numerous and varied. They question the environment's place in today's economy, the appropriateness of focusing on economic growth and, where applicable, its greening. In addition, possible transformation methods such us adapting existing functions, or implementing radical structural changes are considered.

Among these approaches we present the "Transition Studies" which analyse, in a multidisciplinary approach, radical transformation processes of a sustainable society, mechanisms at work, barriers to change, and the levers of transformation.

1.2.1. *The emergence of "Transition Studies"*

Faced with the social issues mentioned above, a number of scientific disciplines have sought to integrate environmental and sustainable development issues. Therefore, new analytical trends appeared in the theoretical corpus of humanities, social and economic sciences. Although these works are interested in the issue of sustainable development and sustainability of the current growth and development patterns, it is essentially in a disciplinary manner in a specific analytical framework that is sometimes amended according to the margin.

If we want to understand what is questioned in the subject of transformation, production and innovation in a transition supporting sustainable development, a more integrative approach is required. It seems illusory to consider that one discipline alone can explain the underlying complexity of the society's transformation processes towards greater sustainability. Fully comprehending what is at stake requires a broad or interdisciplinary understanding, and a systemic vision of the process of transformation and innovation. Standard economic approaches encourage methodological individualism, often from the premise that transitions require above all a change in individual's behaviour induced either coercively (by law, standards) or in an incentive way (through subsidies or by the play of the price system).

In the late 1990s the first works emerged linking the analysis of technological change, socio-technical innovations and sustainable development ("sustainability development"). These works contributed to structuring a new field of research around "Transition Studies". In these systemic approaches, the focus is on the role of institutions and interactions. Thus, "Transition Studies" will consolidate the works centred on the transformation of socio-technical systems [KEM 98, SCH 97, GEE 02, BER 02b], the emergence and dissemination of system innovations [ELZ 04], and the management of complex systems [ROT 00, ROT 01, LOO 07]. This framework for systemic analysis of change is rooted in the traditions of Science and Technology Studies, in evolutionary and institutional economics, in the most recent developments in sociology, including sociology of expectations [ALK 12, BOR 06, VAN 98] and actor-network theory [CAL 86].

1.2.2. *The transition as a process of socio-technical systems transformation*

In "Transition Studies", the concept of transition refers to transformations that occur on a large scale within society or socio-technical subsystems. These changes fundamentally alter the structure of the social system at large. Specifically, Grin *et al.* [GRI 10] propose to characterise the transition processes as co-evolutionary processes that combine multiple changes in socio-technical configurations, and, simultaneously in different dimensions [KEM 05]: technological, organizational, institutional, economic, political, sociological and cultural.

The scope of the introduced changes is considerable and marked by radical changes that contribute to replacing new configurations in an old system. Thus, the premise of "Transition Studies" is that the transition process cannot be based on an adaptation of the existing, or on only incremental innovations that would simply modify the existing system. Echoing the typology proposed by Abernathy and Clark [ABE 85], Geels *et al.* [GEE 04] describe the transition as architectural innovations because they involve substantial changes in terms of both the supply and the user sides. This is the overall architecture or structure of the socio-technical system that is transformed, and the authors stressed that "*without changes from the user side, technological discontinuities are at best characterized as' technological revolutions' that do not include changes in the functionality of systems*". In this context, users are not mere actors nor adopters of new technologies. They will adopt appropriate technologies proposed to co-build innovative solutions. Thus, beyond a literature that studies the development of low environmental impact technologies contributing to the greening of the economy, what is at stake here is not only a shift in trajectory, but the emergence of new technological trajectories able to support a new socio-technical system.

Moreover, the processes of transition are multi-actor and involve a variety of stakeholders (business companies, competitors, suppliers, but also financiers, knowledge producers, universities, public authorities, lobbies, associations, users, etc.), of social groups with different positions and roles, including in the process of creation/dissemination of developed technologies. Some players then take on the role of technology developers, others the role of selectors [BAK 12]. Here the "socio" qualifier makes sense. Innovations that are the bricks of the transition process are not only apprehended under the isolated company prism but as the result of interactions between different social groups.

The socio-technical systems that are analysed in transition processes consist of networks of stakeholders, institutions (that is to say, adhering to standards, whether technical, industrial, legal, and societal as well as practices, customs and values), artefacts and knowledge [GEE 04, MAR 12]. This definition is very similar to that originally proposed by Carlsson and Stankiewicz [CAR 91] about technological systems. Indeed, although forged within separate scientific epistemes, the concepts of technological systems and socio-technical systems refer to the same reality since we consider, as do the evolutionary economists [NEL 82] and institutionalists, that technology is a social construction [BIJ 87].

The various elements of the socio-technical system will interact with one another in order to ensure societal functions to which they are dedicated (e.g. housing, food, mobility, etc.). It is important to remember, as it is the basis of a systemic approach, that understanding the nature of the interactions linking the building blocks of socio-technical systems is central to analyzing the systems of socio-technical system transformation processes and even barriers to which they may be confronted.

The shift from one configuration of a socio-technical system to another results from the processes of socio-technical transitions [GEE 10] that will encompass a set of complementary technological and non-technological innovations in a systemic solution that refers to the concept of system innovation. Elzen *et al.* [ELZ 04] remind us that innovations of systems (which are the innovations at work in socio-technical transitions) generate changes not only in industry, business and technological knowledge, but also in the contexts of use and the symbolic representations attached to the artefacts. Consumer expectations, their view on the services/functions [MON 02] provided or to be provided by the products and technologies, change. These innovations are carriers of functional mutations at the level of society, changing the way to meet the needs and major functions of the society [OEC 10].

Finally, it is clear that the transition processes are expressed at a macroscopic level and are established in the long term. With these processes, new products, processes and services, but also new business models for companies and new organisations will emerge. In other words, technological and non-technological innovations will strengthen themselves, thus creating solutions complementing existing solutions (at least for a time frame during the transition) or aspiring to replace the hitherto dominant solutions.

To understand the contemporary sustainable transition's specifics relative to historical transitions (Box 1.2), it is important to emphasize two points that are actually related. First, and this is what allows us to understand the development of substantial work on the management of the transition and governance, the sustainable transition as presented today is a "guided" transition [VER 12]. Contrary to historical transitions, it does not follow a "natural", spontaneous course of social transformation but it is intentional and goal-oriented (that of reducing environmental and climate impacts of human activities). The sustainable transition involves specific forms of governance [SMI 05]. The regulatory instruments, institutional support and political actors will play a significant role in the guided transition process. Then, the second feature of the sustainable transition element is the pace (accelerated) at which it must be conducted in order to limit anthropogenic effects on the climate and avoid reaching a critical threshold beyond which irreversible phenomena will occur. The various scenarios are well prepared for 2020, 2030 and 2050. At the European level, the energy and climate package adopted in 2008 aims to act against climate change and plans a reduction of 20% of greenhouse gas emissions by 2020, an increase of 20% in the share of renewable energy (hydro, solar, wind, biomass or geothermal) in the total energy consumption of the EU and a 20% decrease in the level of the energy consumption. Revised in 2014, new goals are emphasized for 2030 with at least a 40% reduction in greenhouse gas emissions compared to 1990. The "2050 Energy" roadmap published in 2011 extends these guidelines targeting a reduction of CO_2 emissions by 80–85% in 2050 compared to 1990 levels.

Initially the first work on the socio-technical transitions was designed to analyse the transformation of large technical systems [HUG 69]. In the field of socio-technical analysis, the historical transitions are distinguished from sustainable transitions by their "natural", undetermined, slow character. Among historical transitions we note the movement from sailing ships to steamers in the 19th Century or more recently (1960) the transition from a coal economy to a natural gas economy in the Netherlands [VAN 99]. Geels [GEE 05b] was also interested in the movement from animal traction to the automobile, the distribution of water [GEE 05a] and the establishment of modern hygiene systems [GEE 06b].

Box 1.2. *The "historical" transitions*

1.2.3. *A transition supported by a systemic vision of innovation*

Starting from a technology and innovation-oriented analysis, the different frameworks of sustainable transition adopt a systemic vision of innovation and of transformation processes of technological or socio-technical systems, drawing on institutional and evolutionary approaches [MAR 12]. In doing so, they allow understanding the mechanisms of evolution and transformation of these systems, to identify drivers and barriers to the development and diffusion of environmental innovation or eco-innovation, and to characterize and formalize the process of the transition's management. In this, the authors emphasize "Innovation Studies" which considers that innovation is a process of co-evolution [GEE 02] and that the technical progress contributes only to a part of the answer to the question of what that the future will be.

The purpose here is not to provide an exhaustive review of the literature dealing with sustainable transitions but to focus on concepts and theoretical analyses developed within "Transition Studies". Indeed, economic approaches endogenizing technological change [GRU 02, GRU 04] or favouring the modeling of energy–economy–environment relations in different scenarios limiting greenhouse gases [EDE 06, INT 00) will not be developed here. The same applies for the currents built around corporate social responsibility [POR 06], industrial ecology [SOC 96, EHR 00] or ecological modernization [MOL 00].

Several questions arise concerning the transition process: how does a technological system becomes dominant? How can it stay that way? Why is it challenged? And how do new regimes emerge to question the dominant regime? In a very simplified way, we can consider that the answers to these questions provided by scholarly works in the tradition of evolutionary economics, institutionalism and "Innovation Studies" will focus on the actors, while relative currents of socio-history and "Science and Technology Studies" will further question the methods of construction of a society and the articulation of its components.

1.2.3.1. *Two key concepts: socio-technical systems and niches*

The systemic approach of "Transition Studies" is reflected in the mobilization of two central concepts: the regime (technological or socio-technical) and the niches. The socio-technical or technological regime is a concept shared by evolutionary economics [NEL 82] and socio-history of technologies [BIJ 87]. It refers to the idea that any technology is a social construct. It expresses the socially embodied character of technology, scientific knowledge and influence of institutions (in the sense of rules, norms, customs, values) on the rate and direction of technological change that will develop along technological trajectories [DOS 82]. The emergence of new regimes is characterized by a different way of satisfying societal functions (e.g. energy supply from renewable resources, the proposed soft or shared mobility solutions). Their substitution for dominant regimes characterizes the transition process. The development of new socio-technical systems will be protected from the dominant socio-technical system, in what "Transition Studies" call niches. These niches, whether technological or commercial, are protected areas in which radical innovations will be able to emerge and develop, immune from competitive pressure of the dominant regime. They will be the place of expression of actions of specific public policies, whether from a regulatory point of view, or especially from an economic point of view. Concretely, a niche business model is viable only through public intervention that makes the price of an emerging technology competitive relative to existing alternatives. A mass marketing technology should enable companies to sell their products directly to users without requiring public assistance: a market business model is then characterized by its ability to feed itself.

Ultimately, five lines of analysis dominate "Transition Studies": the multi-level perspective of socio-technical transitions [KEM 94, RIP 95, SCH 94, SCH 96, VAN 87] or *"Multi-level perspective on socio-technical transitions"* (section 1.2.3.2), the *"Strategic niche management"* (section 1.2.3.3), the *"Management transition"* (section 1.2.3.4), the geographical approach to the transition (section 1.2.3.5) and the *"Technological innovation systems"* approach (section 1.2.3.6).

1.2.3.2. *The multi-level perspective*

The multi-level perspective (MLP) sees the transition as a result of the interaction of three levels of a non-hierarchical societal construction: the macro (landscape), intermediate (regime) and the micro level (niches). We clearly find in this analysis the influence of the work of Giddens [GID 84] on models of building a society, the contributions of the classical evolutionary theory of the school of Twente and analysis of socio-technical systems. Following Rip and Kemp [RIP 98] and Kemp *et al.* [KEM 01], different works have the theoretical foundations of this interdisciplinary approach helped strengthen [ROT 01, GEE 02, GEE 04, GEE 05a, GEE 05b, GEE 07, SMI 05, SMI 10, VER 07, GRI 10].

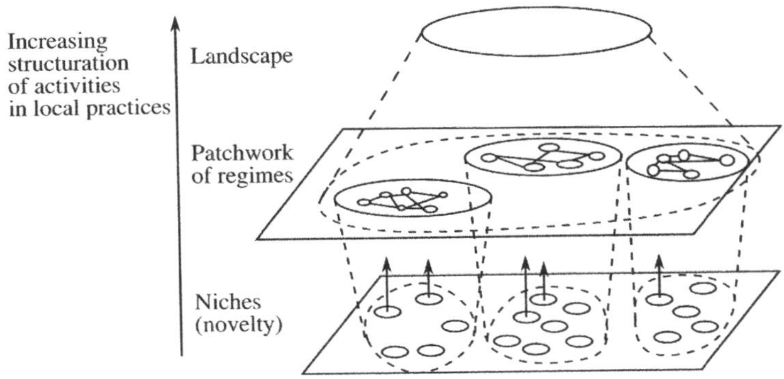

Figure 1.1. *Geels' multi-level perspective (taken from Geels [GEE 06a, p. 173])*

In the socio-technical scenarios that may be proposed [VAN 10], the transition is seen as the result of the meeting top-down oriented flow emanating from the landscape, such as climate change and loss of biodiversity, and bottom-up such as technological developments within niches.

The meso level (intermediate level) is formed by the regimes. Regimes are the set of rules and standards, coordinating the activities of actors and social groups; they provide stability to the system, a certain inertia. They are responsible for lock-in phenomena, which make the actors act according to these rules and them not look outside them [GEE 06a]. The analysis of socio-technical systems requires the study of actors and of the institutionalized relationships they have forged between them. These relationships may constitute barriers to the transition and are being challenged by the changing context and the introduction of new technological solutions and new practices. However, innovations are still visible at this level but they are marginal, incremental or progressive. They grow slowly, making them predictable and attest to the search for forms of adaptation by contributors. The regime is therefore both stable and dynamic [GEE 06a].

The lower level is the micro level, consisting of niches. It is the favored place for the emergence of radical innovations. These innovations may take the form of new technologies, new rules or laws, new organizations, new projects, concepts or ideas [LOO 07]. They emerge in these areas protected from market pressures and competition, and allow social networks supporting innovation to build themselves [GEN 08]. Niches are places where we can deflect from regime rules and where there is the appearance of novelties later tested and transmitted within the regime. The rules are not clearly established in the niches and they offer a space to build them and to experiment. Innovations developed in this way may offend the operating rules and practices of the socio-technical regime [NIL 09]. Their breakthrough in the regime is not guaranteed [GEE 06a].

The higher level is more comprehensive and macroscopic; it is formed of the landscape, which refers to independent aspects of the external environment. It generally affects the development of plans and niches by focusing the changes that occur, given the constraints (e.g. scarcity of certain resources) and objectives (e.g. those of the Kyoto Protocol). The landscape feeds social values and political cultures, and is expressed in terms of socio-political environmental and economic trends. The scenery cannot be changed easily because it does not directly influence the actors [GEE 06a, LOO 07].

1.2.3.3. *Strategic niche management*

The tensions that are expressed between these three levels are in the long-term and changes in socio-technical systems are supported by radical innovations. These radical innovations include technology, but also the organization and the business models of companies. In this sense, a sustainable transition cannot be supported solely by the improvement of existing technologies (e.g. reducing the fuel consumption of internal combustion engine vehicles), or an adaptation of socio-technical systems in place. New regimes should emerge. The dynamics underlying the emergence of these new socio-technical configurations occurs within niches. Their confrontation with the dominant regimes questions the role of the public and the subject of specific developments in the context of the approach to strategic niches management [HOO 02, KEM 98, NIL 09, RAV 05, RAV 10, SMI 07].

Studies of transition in terms of strategic niche management analyse the obstacles to the introduction on the market of sustainable technologies: technological bottlenecks, inadequate regulatory framework, societal values and expectations to meet, insufficient effective demand, no operational infrastructure and lack of confidence in the new technology impacts. Faced with all these factors, we see that the transitions are installed and radical innovations diffuse [WIN 02, GEE 05a, GEE 05b] at two levels: at the level of the global environment and in niches. Technological innovations that emerge in niches diverge from the existing socio-technical system to become part of this scheme.

Beyond the existence of niches and questions on the transformation processes of niches into regimes, empirical work has highlighted the strength of the established actors. Elements of various natures contribute to lock-in a socio-technical system on one or more technologies [ART 88]. The path-dependancy and "lock-in" processes (it may refer to consumption habits, complementary technologies, forms of business model, institutional structures or regulations modes) reinforce the socio-technical systems in place [UNR 00, AHM 08, INT 11]. These systems will then tend to innovate incrementally where the urgency of the transition would require radical innovations combining technological changes and change of uses [MAR 06, FRA 10]. The authors emphasize the need for such radical and systemic innovation to support the development of a sustainable model of society, but also the uncertain and non-deterministic nature of these processes that make sense in the long-term and require special governance [HIS 06].

1.2.3.4. *Management of the transition*

The need to guide the transition towards supporting the development of sustainable socio-technical systems, as well as the need to support the decision process that includes the public, engendered a form of action research (especially in the Netherlands) and reflection on the forms of reflexive governance [VOS 06, VOS 09, NIL 09]. These works are gathered in the approach to management of the transition [ROT 01, KEM 06, LOO 07, KER 08, LOO 10]. Building on the analysis of complex co-evolution of systems process, it proposes problems structuring tools, expression and convergence of antagonistic visions of stakeholders, deployment of agendas and experimentation. Experimentation of new socio-technical systems, particularly their specific use in certain regions, constitutes drivers of change.

1.2.3.5. *The geography of the transition*

It is interesting to note that this questioning joins the level of the most relevant territorial approach to analyse system transformations. It is common to say that the environment is a global problem and that the solutions are to be found at this level. It is on this basis that regulatory processes are sought for major institutional innovations both internationally (entry into force of the Kyoto Protocol) and nationally (promulgation of the "Charter for the

Environment" in 2005, then in 2007, start of the Grenelle Environment process). However, a spatial approach to transition offers an innovative method and gives territories a major role in the transition process. Thus, the geography of the transition [BAT 03, COE 10, COE 12a, COE 12b, RAV 12], aims to reintegrate the influence of spatial and institutional context in the process of transition.

1.2.3.6. *Technological innovation systems*

Finally, in direct line with the work on the technological systems of Carlsson and Stankiewicz [CAR 91], the approach in terms of technological innovation systems (TIS) is to analyse the creative process and diffusion of new technologies [BER 08a, BER 08b, CAR 02, HEK 07, JAC 00, JAC 11]. To do this, it simultaneously considers the various components of this system: actors, networks, institutions and technologies. Markard and Truffer [MAR 08a, MAR 08b] propose defining a TIS as "*a set of actors and institutions networks that interact together in a specific technological field and contribute to the creation, dissemination and use of new technology and alternative/or a new product*". If technological systems initially considered by Carlsson and Stankiewicz are not necessarily innovative, the use of this concept in the context of transition leads to questioning emerging technological systems. The development of this system is an uncertain process of reconfiguration in which the components are led to co-evolve. The authors will then focus on the development, dissemination and use of specific emerging technologies (defined in terms of knowledge and/or product) and talk more about technological innovation systems.

This system aims to fulfill a number of functions or activities that will give meaning and coherence [BER 07a, BER 07b, BER 08a, BER 08b, HEK 07] and failures may justify public intervention. The engine of creation, dissemination and use of technological innovations lies in the systemic interaction that occurs between companies and other actors in a given institutional framework. Several studies have analysed the emergence of sustainable innovations in this context [BER 02a, HEK 07, JAC 04a].

Initially the work in terms of TIS provides a structural reading based on the analysis of elements constituting technological systems, namely the actors (companies, knowledge producers, users, financiers, regulators), institutions, that is to say, rules (codified or not) that help regulate the interactions between actors, and networks that are the links (formal and informal) between the elements of the technological innovation system.

However in order to explain the dynamic process of transformation of socio-technical systems, recent developments in the TIS approach emphasize the need to consider the arrangement of the different activities that will contribute to the emergence, development and the evolution of technological innovation system. It is important to consider that a socio-technical system is not static but evolving. Static analysis of the components of a TIS (or structural analysis) has its limitations since it seeks to understand the mechanisms at work in its constitution. It is important to consider the relationships between the actors (cooperation or not, establishment of barriers to entry for new entrants, technological appropriation, etc.), how they organize their environment or attempt to modify it (lobbying process standardization or normalization, legitimization of technological choices, etc.).

This static view is complemented by a dynamic approach, focusing on the key functions (or activities) that support the development, dissemination and use of new technology [BER 07b, BER 08a, BER 08b, HEK 07, JAC 11, JOH 98, JOH 01]. The main function of any innovation system is to create and disseminate knowledge for the production of innovations and technological change. Six functions were identified [BER 08a, BER 08b, HEK 07], which also contribute to this process. These concern the orientation of the research (encourage actors to invest in a given direction), the development of entrepreneurship (creation of new activities, identifying new opportunities), creating new markets (including stimulating local markets), legitimizing innovations to fight against resistance to change, the mobilization of resources (human, financial, natural) and the development of positive externalities. In addition, these functions are interrelated and can reinforce each other, leading to a process of cumulative causation [JAC 04a, QUI 13].

The development of a TIS does not imply a systematic presence of all of these functions. Indeed, according to Jacobsson and Bergek [JAC 03], the functions may play a different role according to the TIS development. Two key phases of development can be differentiated in particular. The first is characterized by experimentation, in response to strong uncertainties in terms of technologies and markets [KEM 98, VAN 93], when the latter returns to a growth phase, marked by reduced uncertainty and industry consolidation. Furthermore, if the presence of all functions is not a *sine qua non* of the development of an ITS, however, the absence of a function at a given time can be a blocking element or a weakness of the whole system.

The convergence of different approaches, including MLP and TIS approaches, raises questions [MAR 08a, MAR 08b, COE 10], especially from an epistemological perspective. Although the TIS approach tends to neglect the societal pillar of sustainable development for economic and ecological dimensions, it nevertheless has the advantage of retaining the technological systems (including emerging technology systems) as an object of analysis and seeks to explain the mechanisms, including institutional, of emergence and dissemination in relation to their environment. Where the current TIS focuses on the gradual nature of the process of transformation of innovation systems, the current MLP emphasises regime breaks resulting from the deployment of niche technologies. Beyond the recognition of co-evolutionary processes (technology, institutions, organizations), what is at stake is the conception of the thought patterns of a society in construction, and consequently modes of action and governance for sustainable transition.

Let us now see how changes are geared towards sustainability objectives and enabled by eco-innovations in a transition process.

1.3. Eco-innovations: facilitators of the transition?

1.3.1. *Innovation for the environment*

It seems that the term eco-innovation appeared for the first time in the book of Fussler and James [FUS 96], which focuses on the development of value creation processes that take into account environmental impacts. The concept of eco-innovation highlights the entrepreneurial contribution to sustainable development. It can be defined as "*an innovation resulting in a decrease – accidental or intentional – of the environmental impact* [of any entrepreneurial activity]" [OEC 10, p. 15] Eco-innovation is generally defined as "*the production, assimilation or exploitation of a novelty in products, production processes, services or methods of management and business, which aims, throughout the lifecycle to prevent or significantly reduce the risks environmental pollution and other negative impacts arising from the use of resources (including energy)*" [OEC 10, p. 41].

Can these eco-innovations form the basis for the transition? Do they constitute rupture innovations at the base of niches for articulating a radical reform of socio-technical systems, or are they just simple adaptations of the existing regime?

In reality, the concept of eco-innovation has common roots with the classic notion of innovation as developed in the Oslo Manual [EUR 05]. Thus, innovation is different from invention and dissemination and can be applied equally to products, processes, commercial or organizational methods, business practices and external relations. The concept of innovation is understood here as the production, distribution or use of new knowledge with an economic value that will either help improve existing solutions (incremental innovation) or propose disruptive solutions (radical innovations). Eco-innovations are designed in the same way as other types of innovation as new alternatives, which will allow monitoring, limiting, correcting or preventing environmental damage [DEP 09].

Nevertheless, according to the OECD [OEC 10, p. 43], the eco-innovation differs from conventional forms of innovation on two main points: first, innovation is focused on reducing environmental impacts, accidental or intentional; second, eco-innovation goes beyond conventional technological subjects (products, processes, methods) in favour of the inclusion of social and institutional innovations [REN 00]. Thus "*the field of eco-innovation can go beyond traditional organizational boundaries of the innovating company to encompass a broader societal sphere. It thus includes changing social norms, cultural values and institutional structures – in partnership with stakeholders such as competitors, supply chain companies, companies in other sectors, governments, retailers and consumers – to take from the innovation a surplus of environmental benefits* "[OEC 10, p. 43]. In the literature, the concepts of eco-innovation, sustainable innovation, environmental innovation and innovation for sustainable development are often considered interchangeable [CHA 07b].

Beyond this general definition, eco-innovation can be distinguished according to several criteria [AND 08]. This may be its subject (technological innovation: the creation of products and processes, and non-technological innovation: organization, marketing methods, institutional structures, which are new or improved), the mechanisms in place for its implementation (modification, re-design alternatives, creation) or even its environmental impact. However, what the environment takes into account can only be systemic; the field of eco-innovation can't be restricted to incremental technology innovations. It will then affect "*the establishment of new social structures and interactions that involve changes in values and behaviors*" [OEC 10, p. 50] so it is considered that eco-innovation includes the proposed technological and non-technological solutions involving changes in consumer behavior, social norms, cultural values, and formal institutional frameworks.

It is therefore possible to see multiple modes of application of sustainable development principles in companies. This applies to the production and manufacture, but also the tools and management practices, that seek to promote sustainability. Manufacturers will thus find technical solutions to replace toxic materials by non-toxic materials, to reduce their consumption of raw materials and energy, to limit the production of non-recyclable waste. In line with the work of Porter and van der Linde [POR 95] (Box 1.3), three determinants of behaviour adoption of eco-innovations are generally distinguished in the literature: innovation conventional determinants of economy market pull, technology push, and the regulatory aspect.

All these eco-innovations will fuel innovation systems [OEC 10, p. 21] and support a systemic transformation. For the OECD, "*The implementation of the concept of eco-innovation offers hope for the advent of a more sustainable industrial production and support of pressing global challenges such as climate change*".

Porter and van der Linde [POR 95] criticize the traditional approach in environmental economics that considers compliance of companies will only produce additional private costs, that is to say the cost of compliance compared to the objectives laid down in regulation [BON 03]. The heart of their demonstration is to show that well-designed environmental regulations will have the effect of causing, in most cases, innovation on the part of businesses. This innovation will eventually be capable of producing an annuity to cover the costs of compliance and possibly even be a source of additional profit opportunities. The authors seek to recognize, besides the usual incentives for innovation, which are final demand or technical advances, the role of regulation as an incentive factor in the development of these innovations. The authors develop four arguments on the incentive effects of regulation:

– the pollution produced by companies in fact reflects the enterprise resource wastage. Pollution corresponds to a sub-optimal use of these resources and it is rational for firms to substitute other less polluting products;

– as a corollary, regulation can then have a signalling effect for this waste of resources for business, making them aware of the pollution. This was the case, for example, with regard to the changing status of waste from production processes. Until the early 1990s, the waste was only slightly subject to mandatory recycling but through progressively strengthening regulations, companies have realized their potential to use these resources as a source of additional income;

– then, regulations have an uncertainty reduction effect compared to the level of pollution that will be allowed in the sector. It allows companies to engage in investment, but on the condition that defines a sufficient time horizon to ensure the amortization of these investments;

– Finally, defining the acceptable pollution limits, the regulation will have a role in increasing competitive pressure. All companies will have to respect this threshold, companies will want to avoid price competition and will therefore be encouraged to innovate by going beyond regulation to differentiate.

According to Ambec and Lanoie [AMB 09], the adoption of environmental innovations is accompanied with either by lower costs or an increase of income for the company. In the case of looking for opportunities to increase revenue, there are three main transmission mechanisms: first, better access to markets; second, the ability to differentiate the company's products; and finally, the sale of technology to control pollution. The company can play on lowering the cost of: regulatory costs, inputs, resources and energy, capital and finally labour costs.

Box 1.3. *Regulations, an incentive for eco-innovation in companies?*

Beyond the specific aspect that characterizes an eco-innovative product, it is possible to consider this approach in a broad perspective covering new practices of management and innovation.

1.3.2. *New management and innovation practices*

1.3.2.1. *Lifecycle and circular economy*

In this approach, we can first of all consider that a product (or service) is not inscribed in a finite perimeter, but to use a biological analogy, has a lifecycle. A "Cradle to cradle" understanding of this product lifecycle leads to analysing production systems throughout the chain of activities related to the product, that is to say, from the extraction of raw materials, processes manufacturing, distribution, use and end of life management including recycling. Linearly, the economy becomes circular. The circular economy [GAL 16] is "*an economic concept which is part of sustainable development and whose objective is to produce goods and services while limiting the consumption and waste of raw materials, water and sources of energy. This is to deploy a new economy, circular, not linear, based on the principle of 'closing the lifecycle' of products, services, waste, materials, water and the energy*"[4].

1.3.2.2. *Recycling and industrial ecology*

Generally, the analysis of material and energy flows is to maximize the use of resources throughout a given territory (industrial area, agglomeration, region, etc.): wastes of some provide raw materials to others. Recycling is intended to turn waste or recoverable materials into reusable raw materials (this is the case, for example, of textile manufacturing from plastic bottle recycling). This design is the basis for a new industrial ecology or industrial symbiosis in which the industrial ecosystem would become a real vector of sustainable development. This environmental management practice addresses the needs of companies, which, under the pressure of laws, regulations, standards and competition, should integrate environmental strategies [DIE 07]. It faces a quadruple challenge: recycling waste (from "bad" to "good"); complete cycles while minimizing wastes; dematerialize products (increasing resource productivity) and proceed to the decarbonization of energy [DIE 07]. Industrial ecology [FRO 89, ERK 98], ecological engineering or environmental technology and industry recommend to conduct a set of operations of production rationalization (optimization of energy and material consumption, waste minimization at the source, reuse of waste for use as raw materials for other production processes).

The idea that the product has a lifecycle finds its extension in a second idea which is that the same product may, subject to refresh or repair, have a second life as such [HEY 14]. It is then necessary to imagine the successive multiple uses of a product, from first life, then to second life until dismantling for recycling. For example, in the field of electromobility, one can imagine that the batteries used today to power electrical vehicles (first life) can find a second life for use in electricity home management systems. The circular economy will thus promote not only the repair of goods (availability of spare parts, extending the statutory warranty period), but also re-use as part of second-hand chains.

---

4 http://www.developpement-durable.gouv.fr/L-economie-circulaire,45403.html.

These approaches directly impact the product design process as it takes place upstream of the value chain. The eco-design approach is based on the concept of product lifecycle and the idea of the need to integrate the various stages of the product life, from birth to end of life, from its design in order to limit the environmental impact. In this context also, the evaluation of the environmental impact of a product will have to take into account all the stages of all the elements of the value chain.

1.3.2.3. *Functional economy*

A third element is to reflect on the meaning of the products. This is why we manufacture them and their conditions of use. Prioritizing use over possession of a service or a good is the aim of the functional economy [BOU 14]. It provides a framework of business activity analysis centred on the ability to offer a service, not a product to address an identified feature. One of the first companies to have initiated this approach was Xerox, offering rental services of photocopiers. Similarly a company like Michelin who proposes to sell to the transport industry, not the tires but the kilometers driven is part of this approach.

In recent times, the functional economy has integrated sustainability issues and co-production solutions [DU 11]. It may be noted that the different views that companies can have on their products, processes and more broadly on the nature of their activity, are also found at the consumer level. Thus, we see the emergence of engaged consumption, responsible citizen trends, characterized by a different relationship with the product. These movements are a realization within various associations as discussed in the case of food, and it is also the philosophy that is found in DIY trends (Do It Yourself) or in Fab Labs [MOR 16]. Responsible consumption leads the buyer, whether economic actor (private or public) or citizen to make his choice, taking into account the environmental impacts at all stages of the product lifecycle (goods or services) but also fighting against waste and planned obsolescence, expressing a need for traceability.

In France the circular economy participates in the law on energy transition (Box 1.4), but it is also an internationally widely shared concern. Today, many countries are explicitly committed to supporting the circular economy (Japan, Germany, the Netherlands, China). They coordinate or encourage action at all levels of government (cities and municipalities, neighbourhoods, districts, regions, States) [ROU 14]. The lack of space and/or resources seems to be a common feature of the circular economy pioneer countries. To this first motivation are added concerns of national independence, defence of economic interests and the fight against exports of strategic items that are also central. The policies of these countries in terms of the circular economy are established and the role of the state is conceived as that of a catalyst to remove administrative and regulatory brakes as part of public/private cooperation. Japan has a framework law, a law on the promotion of efficient use of resources, a law on waste management, a law on the promotion of green procurement and laws that are sector-specific. In Germany, the decoupling of growth with material consumption is registered in the national sustainable development strategy. China has a law on the promotion of circular economy, inspired by German or Japanese devices, exceeding the only fields of energy or waste to deal with all resources. If forms of circularity are most upstream (eco-design, repair, re-use and re-use of equipment) only Japan promotes the design of easily recyclable products, labelling for recycling, the use of co-products, etc. Other countries may disclose less advanced approaches, where the term circular economy does not appear, but with sector results quality (Finland, Brazil, etc.). Beyond these institutional aspects in the most advanced countries such as Japan, Germany and the Netherlands, many achievements are based on voluntary commitments, possibly accompanied by tax incentives and subsidies.

Research on transitions and sustainable development, whose guidelines have just been redrawn, is developing rapidly to respond to the need to better understand the corporate transformation process and questioning the process of emergence of new technologies and new practices likely to contribute to the sustainable society. This chapter led us to propose a new concept, that of techno-ecological transition that considers the need for a structural transformation of societies integrating the environment and operated by a technology seen as a social construct. Considered from a systemic perspective, techno-ecological transition refers to changing technological or socio-technical systems centred on the inclusion of environmental innovations and technologies being thought of as places of knowledge crystallization, expectations, actor practices and finally values or imaginaries.

The objective is to move from an economy of predation of resources – natural and human – to an economy of conservation and moderation of these resources. While the ecological transition focuses on the relationship between man, society and the environment, nature and the socio-technical transition focuses on human interaction/technology in the context of environmental constraint. The concept techno-ecological transition focuses, in turn, on systemic environmental innovation and eco-innovations and the dialectic between technological systems (in their tangible and intangible aspects) and the environment, as we shall see in the next two chapters that will address the techno-ecological transition in the areas of energy and agriculture-food.

> "Art. L. 110-1-1. – "The transition towards a circular economy aims to exceed the linear economic model of extracting, manufacturing, using and wasting, calling for a sober and responsible consumption of natural resources and primary commodities and, in order of priority and prevention of waste, including the reuse of products, and according to the hierarchy of waste treatment methods, to reuse, recycle or, alternatively, to recover wastes. The promotion of industrial and territorial ecology and ecological product design, use of materials from renewable natural resources sustainably managed and produced by recycling, sustainable public procurement, the lengthening of the duration of the lifecycle products, waste prevention, prevention, reduction or control of the release, the release of the flow or the emission of pollutants and toxic substances, waste treatment respecting the hierarchy of treatment methods, cooperation between economic actors at the relevant territorial level in accordance with the proximity principle and the development of values of use and sharing, and information about their environmental costs, economic and social help the new prosperity".
>
> "Art. L. 110-1-2. – "The provisions of this code are intended, primarily, to prevent the use of resources and to promote a sober and responsible consumption of resources and to ensure a hierarchy in the use of resources, focusing on resources from recycling or renewable sources and recyclable resources and other resources, taking into account the overall balance of their lifecycle."

Box 1.4. *The circular economy in the French law on the energy transition for green growth*

2

Energy Transitions

The challenges of climate change and energy security in the context of overall growth in energy demand calls for radical changes in energy systems [FOX 08]. These changes will involve considerable financial investments, technology choices and modifications in the behaviour and expectations of consumers. In this context, the energy transition refers to the set of structural changes of the current energy system, changes that are necessary to facilitate the transition from a society based on the consumption of fossil fuels (coal, oil, gas, uranium), strongly emitting greenhouse effect gases (excluding uranium), to a society that is more energy-efficient, incorporating and increasing the share of renewable energy in its energy mix. This concept simultaneously covers the evolutions of the energy forms, modes of production of the energy used and particularly the decarbonization of electricity, as well as how this energy can be used in the main sectors of economic activities, including transport.

This issue of energy transition holds a dominant position in the discussion of ecological transition to which it is sometimes assimilated, especially in France. For some, this transition is the basis for the so-called third industrial revolution [RIF 12]. The issue of energy transition has been the subject of numerous discussions on a national scale, with conferences organized in France from November 2012 to July 2013[1] which led to the

1 See http://www.developpement-durable.gouv.fr/IMG/pdf/dnte_synthese_web_bat_28-8.pdf for a summary of the debates.

law of energy transition for green growth, published in the *Official Journal* on 18 August 2015[2]. In this context, several quantitative targets have been set. The goal is to reduce GHG emissions by 40% in 2030 compared to 1990, to reduce the consumption of fossil energy by 30% in 2030 compared to 2012, to increase the share of renewable energies of the final consumption of energy to 32% in 2030 and to 40% of the production electricity, to reduce final energy consumption by 50% in 2050 compared to 2012, to diversify electricity generation and reduce by 50% the share of nuclear in 2015. This quantitative approach echoes the proposals from prospective energy scenarios which compare emissions of CO_2, changes in energy prices, the degree of security of supplies, necessary investments and the impact on employment [CHY 15].

The interest expressed by economic and political actors and the multiplicity of recent works and reports [INT 08b, INT 13, INT 15, LE 15, SDS 15] which have stimulated reflection and debate, indicate that we are up for the challenges. It reflects the central place that energy occupies in our society and its structuring nature. It is at the core of economic and social activities, to the extent that the issue of energy transitions spans technological, ecological, economic and social issues.

The objective of this chapter is to highlight what is covered today by the notion of energy transition, understood as a transformation towards a low carbon energy production. We are interested therefore in what constitutes the current energy system, in reconfigurations that emerge in the energy sector marked by the development of new energy technologies and the evolution of responses to energy needs. This vast subject is explored in five sections. The energy system appears as a historically and socially structuring item (section 2.1). If one accepts that the concept of energy transition refers to a set of changes to be made in the current energy system to make it more sustainable (and this for multiple reasons that will be explained below), it is important to characterize this energy system (section 2.2) and the limits to which it is confronted today (section 2.3). Facing these limits, a range of technological solutions can be combined to support the transformation of the energy system (section 2.4) but they also face socio-economic and institutional barriers (section 2.5).

2 https://www.legifrance.gouv.fr/affichLoiPubliee.do?idDocument=JORFDOLE000029310724 &type=general&legislature=14; Texte: http://www.assemblee-nationale.fr/14/pdf/ta/ta0575.pdf.

2.1. A socially structuring energy model

Despite its importance, understanding what is energy is not easy, so it is an abstract concept. The term "energy" appeared in the 19th Century and its first meaning refers to "what is contained in action, in the work", and "that which allows a change in the state of things". Understandably, energy does not exist as such but takes many forms [SOL 10]. It can be neither created nor destroyed, in accordance with the principle of conversion and its measurement will vary according to the objectives (Box 2.1).

The measurement of energy relies on various conventionally established units.

Joule: a joule is equal to the work done by a force of 1 N

Watt-hour: measures the energy power available for 1 h

TEP: represents the raw energy provided by the complete combustion of 1 ton of oil (or 11.6 MWh by convention a calorific value of 42 gigajoules – GJ)

Box 2.1. *Measuring energy*

Over the centuries, man has gradually learned how to domesticate sources of energy through the development of science and technology. Being able to transform our environment by mobilizing this energy is a first step. This permanent conquest on nature began with the control of combustion, then with photosynthesis, mechanical energy and more recently electricity and the atom [JAN 13]. The ability to mobilize more efficient energy sources contributed to the improvement of living conditions in the space of a few decades. What was initially a slow process of transformation of societies, witnessed an extraordinary evolution in the 20th Century [OEC 11]: the growth of economies intensified (multiplication by 22 of the world's wealth), as did population growth (multiplication by 4 of the world's population estimated at nine billion people in 2015). Fossil fuels, whose consumption was multiplied by 14 during the 20th Century, are the catalyst of this acceleration.

At each of these stages, a re-structuring of society and the economy led to the current configuration of socio-technical systems in which fossil energy is a key component. The models of society are thus crystallized around these energy resources [SCH 12]. As noted by Rumpala [RUM 13], "*the techniques developed and infrastructures deployed materialized our current modes of collective organizations* [...] *engaging users in more or less strong value chains*". While previous energy transitions are made at the rate of the deployment of an alternative technology offer supported by the innovative entrepreneurs such as described by Schumpeter [SCH 35], allowing socio-technical systems to rebuild around new arrangements, the current transition is more the translation of environmental and societal constraints that production must integrate. This "conceptual inversion" [SID 13] leads us to question the role of innovation in this process of energy transition. There is no doubt that the current energy transition is part of the current climate emergency.

By basing economic development on an exhaustible and polluting energy source, society is now confronted with the limits of its development. In the past, energy transitions have already occurred [DUR 14b]. Today, energy systems must be transformed again to reflect the different challenges they face, but the implementation of a "low carbon" society goes hand in hand with a change in the relationship with the environment and nature. In doing so, this involves, through a systematic innovation process, a paradigm shift in how we understand energy and its contribution to economic activities. It is the role of the visions proposed by different prospective energy scenarios that tries to construct possible futures. For instance, the General Commission of Sustainable Development [COM 15b, pp. 9–14] proposed three vision of society: "green civilization" integrating the ecological constraint in a search for green growth, supported by enlightened consumerism, a tripartite governance; "biocivilization" based on a paradigm shift organized around environmental realism, of a social equity research (cooperation, solidarity, responsibility) and bottom-up governance; "connected civilization" combining technological democracy, new humanism, an economy of sharing and functionality.

The question of energy is, therefore, extraordinarily rich and complex, suggesting what might be the contours of tomorrow's society and what suggests systems' innovations that emerge within major societal functions and their contribution to energy transition. The transformation of our energy system represents a change that exceeds the field of technology to cover socio-technical but also economic and environmental developments. It invites us to try to better understand the impact of our lifestyles on the environment, to look out for signals of energy insecurity, to reflect on what our fundamental needs are, how to satisfy them today and how they can be satisfied tomorrow. This is a unique opportunity to question our socio-technical choices and build a society that provides food and shelter to everyone with a total respect for human integrity. As recalled by Chevalier *et al.* "*[...] as soon as we put our mode of energy consumption 'on the table' to dissect it, we inevitably emerge with fundamental questions concerning the model of society in which we want to live* [...]. *The energy transition debate automatically leads to thinking of new forms of growth*" [CHE 13, p. 112].

This shift in paradigm should not only be a political project for the reconfiguration of energy systems and the reallocation of public investments, nor can it be addressed only from the point of view of technological solutions or the engineering of energy production. We need more than scientific knowledge and technological progress alone to provide the answers to the many challenges that we currently face.

2.2. Fundamentals and characterization of the current energy system

2.2.1. *Energy and the energy system*

Defining an energy system requires a techno-economic approach. Within this framework, we will define an energy system as the articulation of various forms and sources of energy (i.e. raw materials or natural phenomena used to produce energy) mobilized in an economic system, whether local, national or global [HAN 10, DEB 13]. Any energy system builds on natural resources that may be finite (stock such as carbon) or infinite (stream as solar radiation), renewable (wood) or not (petroleum).

To understand the economic function of this system, note that, from the classical physics point of view, energy allows the transformation of matter in different states. It has the ability to modify the state of the systems with which it interacts, which is manifested by the production of heat (thermal energy in heat engines with internal combustion), mechanical movements, radiation (light), by fission or splitting of atoms (nuclear energy) or by chemical processes (chemistry of the fuel cell). In this way, energy appears as heat or cold, as light, as transformation of matter, which leads it to play a central economic role in activities of production.

It is by learning how to mobilize energy's transformative force, regardless of its shape and source, that man has shaped his environment over time. Energy mobilized in the framework of economic activities will generate growth and economic development, and these will be correspondingly higher than the resources mobilized will be efficient and will multiply human strength. Therefore, it is understandable that a society's energy consumption is indicative of its ability to configure the world surrounding it, to build its environment.

The human relationship with energy has evolved. The first act of this process was certainly the domestication of fire some 500,000 years ago, which allowed the settlement and creation of social and institutional structures. For centuries, man has used essentially renewable resources from nature (water, animal strength, wood/biomass) and natural phenomena (solar radiation, wind speed). Then, non-renewable energy resources, drawn from the bowels of earth and formed by the slow decomposition of organic materials, were added. These resources are fossil fuels. Coal, used for over 5,000 years in China, was the essential resource of the First Industrial Revolution, along with the steam engine. The first petroleum wells appeared in the United States, in Pennsylvania, in 1859 and then a couple of years later in Azerbaijan (1870). Today, global energy consumption is largely powered by these hydrocarbons: coal, petroleum and natural gas, which make up the vast majority of the world supply of said primary energy.

2.2.2. *From primary energy to final energy*

Primary energy refers to energy available in a raw state on the planet and accessible before transformation. It exists in two main forms. First of all, it comes in the form of raw materials or resources stored in the different layers of the terrestrial subsurface (coal, crude oil, natural gas, uranium, unconventional gas). These energies are inherently non-renewable or exhaustible taking into account the time necessary for their production. The three of them – petroleum, coal, and natural gas – satisfy 80% of the world's energy needs. Dominant until the end of the 1960s, coal represents today a quarter of the primary energy used worldwide, oil accounts for 30% and natural gas for 20%. Then, energy was also made accessible for us in the form of flows related to natural phenomena, which are perennial but intermittent (solar, wind, hydraulic energy). These sources are called renewable in that they have their own capacity of reproduction, independent of all human intervention. These primary energies will undergo various transformations and power the energy system by becoming secondary energies that may be carried and used as final energy. Secondary energies include petroleum products, solid and gaseous fuels, and electricity[3].

As for the energy that we use daily (electricity, butane, gasoline, fuel oil, coal) in order to satisfy our various needs, they are called final energies. Final energy is not available readily in nature. They are therefore the result of the transformation of primary and secondary energies (for example, the refining process that turns crude oil into gasoline or fuel oil), a transformation during which energy loss occurs. This distinction between primary and final energy covers a distinction more familiar to economists between energy production and consumption. Typically, primary energy consumption is greater than final energy consumption. The difference representing losses linked with the energy conversion and the energy efficiency of the various technologies is used for the production and use of energy.

The current energy system is built according to a process of aggregation of primary energies. One or more resource types can be dominant at any one time and substitutions are made gradually. The weight of these various sources of energy in the energy system varies from one country to another, depending on the resources present and exploitable in its territory, the evolution of its needs in relation to its level of industrial development, its demographic growth, the energy technologies available and usable and energy policies implemented by the states.

To appreciate the weight and role of the energy sector in the economy of a country, it is customary to report the total energy consumption in this country to the created wealth, measured by the gross domestic product. This indicator highlights the spatial and temporal disparities that can be explained by multiple factors in the study area: its geography along with its climate, the extent of its territory, its productive and economic structure, which impacts the nature of the energy used, its technologies, and the greater or

3 Electricity has a particular status which we shall return to later. Like hydrogen, it must be produced by a primary energy conversion.

lesser efficiency of technical systems producing, transforming and using energy. The expression from an accounting point of view of an energy system appears through what is called its energy balance. It reflects the way in which, during a given period (generally a year), a country will produce, import, transform and use energy[4]. The energy balance thus allows recording flows of energy consumed by a country and identifying their origin. The different sources and forms of energy and expressed according to their calorific value and conversion factors are introduced, which are a subject of controversy, particularly with regard to electricity [HAN 10, pp. 3–5]. It is the structure of this energy balance that transition energy aims to modify in order to reduce the negative consequences of heavy usage of fossil fuels.

In this type of description, the energy system remains largely disembodied. If we accept that a system is driven by functions to be completed but also by the game of the actors in presence, of artefacts and institutions they build, it is possible to have a different reading of the energy system. Without this grid that allows updating the processing mechanisms and assessing the impact of the mobilized levers, it is difficult to understand the sticking points that affect the process of transition of the energy system. The approach by technological innovation systems allows us to identify all of the actors involved, the links between them in networks supervised by institutions and infrastructures (systems of knowledge, of communication, systems of transport and energy, market structures, control systems, of standardization) that interact in the production, dissemination and use of technologies [JAC 00, JOH 01, SUU 09a].

2.3. The limits of the current energy system

The world's energy system has been built around the use of fossil fuels and has now reached its limits. The different scenarios built around transition energy highlight the unsustainability of the current situation. The analysis of its limitations, the controversies arising therefrom (e.g. around the "Peak Oil") highlight a real assessment problem of the state regarding the energy system[5]. This issue is even more crucial than the construction of

4 The functioning of modern economies involves continuous flows of energy of which a part is imported. For oil- and gas-importing countries, energy dependence and security of energy supply are the two main concerns. These points are not developed here.

5 Klitkou and Borup [KLI 12] propose to establish a follow-up of the evolution of the systems of technological innovation in energy through a set of statistical indicators.

energy scenarios built on a state of quantification and assumptions, which may be more than expected evolution. In a search for technological solutions to reduce the anthropogenic effects of human activities, an abundant amount of literature aims to model the process of mobilization of low-carbon technologies. Two major challenges must be considered: the shortage announcement of fossil fuels (oil and gas) and climate change linked to CO_2 emissions. The production and consumption of energy on an industrial scale automatically entails a tension on the environment related to the link between growth and consumption of energy. If this tension has been absorbed so far, it is less and less the case in the current demographic and industrial context. However, it seems that the rate of the transition and the incentives to change the system are dependent of the ability to assess the available fossil reserves, and to assess the volume and impact of CO_2 emissions and more generally of greenhouse gases.

2.3.1. *An evolution of the world's energy consumption...*

The depletion of energy resources is linked to a significant increase in demand/consumption in the world. Indeed, the global primary energy demand has experienced an explosion in recent decades, in relation to population and economic growth worldwide. These developments are accompanied by an increase of the unit energy consumption directly linked to the ways of life that have been developed particularly since the first Industrial Revolution. Thus, not only has the population increased substantially but the energy consumption per capita has also greatly increased especially in emerging countries (China and India), which are developing industrial and residential sectors. Between 1973 and 2012, the global production of primary energy has more than doubled to settle at 13,371 Mtoe in 2012. The global primal energy mix thus consists of 82% of fossil carbons (75% for the EU), 13% of renewable energies and 5% of electricity of nuclear origin [COM 15c].

For economists, energy is both a final commodity used by households for heating, lighting, moving, operating equipment of daily life as well as a good of intermediate consumption leveraged by businesses in productive processes to produce goods and services: "*Manufacturing industries represent a non-negligible share of global consumption of resources and production of waste. Globally, their energy consumption has grown by 61% between 1971 and 2004, and is almost a third of global consumption. These* *industries are responsible for 36% of global emissions of carbon dioxide (CO_2)*" [OEC 10, p. 22]. A part of this energy consumption is designed to meet societal needs while being fairly closely dependent on the terms of economic growth and the structure of the GDP. Energy consumption, and the manner of this consumption, is closely linked to the level of development and growth of a country, which its citizens see as part of their fundamental needs. We can see here how the questioning of the current energy system led to a strong societal questioning on the needs to be met once one accepts our inability to maintain a way of life and production that is so energy-intensive. As such, we note that although the development of a digital economy initially received a green image, studies on environmental impact of information technologies and of communication show that it is far from being the case [FLI 09, AIT 10].

2.3.2. *...which shrinks fossil energy reserves...*

The current rate of exploitation of exhaustible resources, particularly oil, coal and natural gases, has led to a massive destocking without possibility of considering a natural recovery of this stock, even after several generations. In fact, our current reserves are dwindling considerably. One of the key debates of environmental economics concerns the estimation of the available reserves (or stock) of coal, oil and natural gas.

This notion of reserve has been progressively refined over time, resulting in a normalization of the concept. So when mentioning the concept of a reserve, it is not so much to opt for an absolute physical estimate, as to favour a physico-technico-economic approach in which economically bearable costs play a key role. We will distinguish proved reserves (updated and exploited), probable reserves (discovered but unexploited), inferred reserves (undiscovered but with a supposed existence on existing exploitation sites) and speculative reserves (involving undiscovered reserves on unexplored sites). Globally proven reserves of fossil resources are published annually in the *Oil and Gas* journal and in *World Oil*. Their volume depends on the cost that companies will consider if the resource is exploitable and economically cost-effective. We estimate these reserves have up to 40 years for oil, between 60 and 100 years for coal and 180–200 years for natural gas.

But these estimations are never final and should be periodically reviewed. Indeed, the increase in reserves may result from four distinct elements: the discovery of new reservoirs (currently the discovery of new reservoirs does not compensate for the growth of global energy consumption), including deposits of non-conventional gases such as shale gas; the improvement of the recovery rate of fossil resources during their extraction from the subsurface (advances in technologies of extraction of fossil resources allowed to pass a 25% recovery rate in the 1980s at a current rate of 35%, which is equivalent to an increase of 50% of proven reserves); the changing of economic conditions (the increase in the price of a barrel of crude oil increases the profitability of some most difficult reservoirs in access, therefore more expensive); the strategic behaviour of actors (since the value of the shares listed on the oil stock exchange is proportionate to the amount of reserves that they say they have).

The result of the exhaustion of oil reserves (but this is, of course, valid for coal and natural gas) is that it theoretically generates an increase in the price of this resource (Box 2.2). This situation is expected to stimulate the search for alternatives, substitutes and new reservoirs (increase of proven reserves), but the difficulty of assessing the available stocks will in fact introduce uncertainty about the duration of possible exploitation of these resources. However, this conditions the rate in which the energetic substitution could be done and impacts the price of this resource. Therefore, allowing the transformation of the energy system to be done "naturally" through the laws of the market and prices regulation can only lead to an impasse. Many actors point to the fact that it is impossible to rely on the rapid depletion of fossil resources to reduce CO_2 emissions.

Harold Hotelling, in *The Economics of Exhaustible Resources,* [HOT 31], seeks to build an indicator of economic scarcity that will apply to exhaustible resources. In economics, Hotelling's annuity or annuity of scarcity is the annuities of the owner of a stock of non-renewable resources. It refers to the difference between the marginal cost of production of a non-renewable resource and the market price. The more the resource becomes scarce, the more the annuity increases. The maximal annuity implies the exhaustion of the resource; the price is so high that its demand is void.

According to Hotelling, in a situation of free competition where the quantity of reserves is known in advance, the annuity of the non-renewable resource grows at the pace of the interest rate because the owner of a stock of resources makes an arbitration between selling now or selling later.

Box 2.2. *Hotelling's annuity*

2.3.3. ...and generates local and global environmental damage

Beyond the greater or lesser availability of fossil resources, their destocking generates an even greater environmental nuisance as their exploitation intensifies. These environmental nuisances are of two types: firstly, the deterioration of local ecosystems (local pollution of extraction sites, various risks as in the case of shale gas), and secondly, the broader effects related to the release of greenhouse gases when using fossil fuels whose combustion will generate harmful gases. From an economic point of view, the internalization of environmental impacts involves their measurement, especially when accounting for GHG emissions [RAR 11].

Economic analysis is, however, less concerned with the consequences on economies and local ecosystems of the exploitation of fossil resources and global warming. On the occasion of the COP21, Woesner [WOE 15] offers an eclectic panorama of the current situation. Global environmental impacts were widely discussed, especially as regards global warming [TRO 09]. As indicated in Chapter 1, a substantial part of the warming has its origins in greenhouse (especially CO_2) gas emissions related to human activities. CO_2 emissions come mainly from the burning of fossil fuels (coal, natural gas, oil), in the transport, residential-tertiary (building) and industry sectors.

According to estimates from the International Energy Agency (IEA), today, 80% of CO_2 emissions worldwide are due to the combustion of fossil fuels – the last 20% comes from five other greenhouse gases considered by the Kyoto Protocol: CH_4, N_2O, PFCS, HFCs and SF_6 (Box 2.3). This combustion is implemented both during the process of transformation of fossil fuels (petroleum refining, electricity) and end uses of these energies (in transport, housing, industrial activities). In recent years (between 1990 – year taken in reference by the Kyoto Protocol – and 2012), the IEA estimates that global CO_2 emissions from combustion of energy increased by 51%, thus reaching 31.7 billion tons of CO_2 issued [COM 15a, COM 15b, COM 15c]. If correlated to GDP, these programs tend to decrease in most of the countries; on the other hand, correlated to the population, they stagnate. CO_2 emissions related to fossil energy per capita consumption still present significant variations depending on the country. According to the IEA [INT 14], they settle 6.1 t CO_2 per capita for China (with a multiplication by 3 in 20 years), 5.1 t CO_2 per capita for France and 16.1 t CO_2 per capita for the United States (with a strong decrease since 2005 when the level of CO_2 per capita reached 20 tons).

> "*At the request of the United Nations (UN), the Intergovernmental Panel on Climate Change (IPCC) has proposed two methods of inventory relating to the CO_2 emissions from the combustion of fossil fuels (IPCC Guidelines for Greenhouse Gas Inventories). The first method, called the 'sectorial approach', estimates CO_2 emissions from the consumption of energy for energy use, distinguished by sector (residential-tertiary sector buildings, industry, transportation, agriculture, power generation, petroleum refining, etc.). The second method, called the 'approach of reference', also uses energy consumption, but the distinction of their energy use (and therefore CO_2 emitter) or not, is made simply from allocation keys. It is therefore less specific than the first method and does not allocate emissions between sectors.*" [COM 15a].

Box 2.3. *Methodology of CO_2 emissions evaluation*

Ultimately, these elements lead us to emphasize the need for decoupling between the consumption of energy, economic growth and population growth [AND 05]. A new phase begins, characterized by questions regarding the possibilities of green growth [OEC 10, OEC 11], using less energy while being creative in jobs. In being interested in developments in consumption and prices of energy resources, a whole field of the economic analysis is developed within an economy of the environment. One of the hypotheses put forward is that the awards are the necessary information in decisions of the actors. The price system appears here in its traditional function of signalling to guide the investment behaviour of firms and the consumption of households [ALB 09] and to link public actors and markets. The market regulation is necessary and sufficient.

As recalled by the OECD, "*the uncertainty relating to the economic cost of traditional energy resources and the need to combat the local pollution and the effects of climate change are strong incentives to diversify energy systems and to radically improve the effectiveness of the production, storage, distribution and consumption of energy*" [PER 15, p. 23]. This generates two questions: that of the rate of this transformation, which must necessarily be accelerated in the context of the 2°C Scenario which implies that at the global level the GDP energy intensity and carbon intensity of primary energy need to both be reduced by approximately 60% by 2050 [INT 15]; the cost of the GHG reduction, which "*could not be maintained at a reasonable level without appeal to a portfolio of technologies ranging well beyond that currently available*" [AGH 09]. What is the role of the technological innovation in this process?

In order to foster the energy transition, define the energy mix and guide technological trajectories, actors have imagined different scenarios of transition. But an oriented-scenario approach leads to focus on an objective (environmental, on the one hand, and on the configuration of the energy mix, on the other) and tends to ignore the conditions to orchestrate this transition. This path is constructed incrementally as we can see in these "transition studies".

2.4. Innovation in the energy transition

The so-called new technologies of energy are at the heart of the transition: thinking of innovation in the field of energy is indispensable [SID 13]. It is clear that new energy technologies can contribute to the mitigation of climate change and that supporting (including through public investment) the development of new energy technologies can accelerate the rate of the transition to a low-carbon energy system. However, as stated previously, the relationship between innovation and transition needs to be reconsidered. The current process contrasts with those of the past where energy transitions were generally able to be done at the rate of the development of alternative technological solutions.

To implement the energy transition and create a decarbonized energy system, three main technological orientations exist, from the adaptation of current energy technologies to the proposal of solutions of rupture, through a diversification of the energy mix[6]. The deep decarbonization of energy systems[7] may be orchestrated around energy efficiency, the decarbonization of energy vectors (especially electric) and the development of low carbon energy vectors in the end uses of energy such as transport.

6 The concept of energy-mix reflects the combination of the primary resources that mobilizes a country to meet its energy needs.

7 The "deep decarbonization pathways" project launched in 2013 by J. Sachs, economic advisor of Ban Ki-Moon, aimed at the production of scenarios of decarbonization of the energy systems of 15 countries. It was relayed in France by NACES (National Alliance Coordination for energy research) in order to identify the technological levers. "Decarbonization wedges" allowing to significantly limit CO_2 emissions in the long term.

2.4.1. *Improving the current system of energy production and adapting existing technologies to new constraints*

Mitigating the negative effects of current patterns of production and consumption, scaling existing energy technologies, making them more efficient and less polluting by regulating combustion emissions, are all solutions for innovatively supporting the energy transition.

2.4.1.1. *Producing energy more efficiently*

This concerns the development of solutions to improve the modes of production, transportation and distribution of energy. Indeed, during the various stages of conversion of energy, a reduction in yield energy appears. We must implement technological innovations to increase energy efficiency without generating a rebound effect (Box 2.4). Energy efficiency aims to maintain the level of current energy services (services to ensure mobility, heat, etc.) while reducing the drain on the primary energy including fossil resources.

In 1865, while studying the question of the future of a coal-based economy, Jevons noted that technological advances introduced by James Watt with the steam engine helped to increase the profitability of coal energy, which, in turn, allowed Watt's technology to be more useful than that of Thomas Newcomen, thus increasing the consumption of coal.

Jevons' paradox is now known as the fact that the most efficient usage of an energy resource leads mostly to the consumption of this resource and not a reduction. Indeed, by improving the use of a resource, we reduce its price. This also allows for the emergence of new uses. In this case, the efficiency effect is more than offset by a negative volume effect. The rebound effect helps to explain why, despite the important progress made in terms of consumption of combustion engines for several decades, gasoline consumption continues to grow in industrialized countries [MIS 14].

Box 2.4. *Rebound effect or Jevons' paradox*

2.4.1.2. *Consuming energy differently*

A second field analyzed in the literature concerns the improvement of energy efficiency in its different uses. Here "*the concept of efficiency refers to the 'technical' improvement of the systems in place to consume less energy without degrading the quality of the service provided (especially at home). It is as well to improve the performance of infrastructure and objects consuming energy (plants to refrigerators, also cars) to rethink their trajectory (i.e. including the recovery and recycling of the materials used). Sobriety more fundamentally challenges our ability to moderation to restrict energy purposes that are often identified as mismanagement up to now*" [CHE 13, p. 112]. This approach could contribute to halving greenhouse gas emissions in the main consumption sectors of construction, transport and industry.

As such, in France the building sector is one of the highlights of energy transition projects. The use of new materials for thermal rehabilitation of buildings and the construction of buildings with positive energy[8] are actions listed in the law. Similarly, the transport sector is also heavily affected. It is often presented as one of the activities with strong environmental impact, both at the level of pressure on fossil resources that it requires, as well as at the level of emission of polluting particles that it induces and of which we know the effects in terms of climate change and human health. In following regulatory obligations, car manufacturers have often innovated to reduce the harmful effects generated by cars and trucks. Examples of regulatory obligations include the standards on emissions of CO_2 and of particles that led to the establishment of particulate filters in car engines, or the recycling standards requiring manufacturers to consider the entire lifecycle of vehicles and to consider the materials used. However, in the absence of deep behavioural changes, these advances are thwarted by increases in traffic volume. In other words, gains in efficiency essentially allow for some mitigation in the acceleration of environmental degradation.

2.4.1.3. *Sequestering carbon*

In a move to mitigate the negative effects of human activities and correct the nuisances generated by our current energy systems, another method involves sequestering (capture and stock) the CO_2 released during industrial production processes [INT 08a]. Indeed, certain industrial activities are particularly polluting and CO_2-emitting. In order to limit the presence of CO_2 in the atmosphere the objective is to capture it (at the exit of plants) and then

8 A building that is able to produce more energy than it uses.

store it. The capture and geological storage[9] of CO_2, Carbon Capture and Storage Technologies (CCS), is a technological solution for the decarbonization of the economy: "*to capture CO_2 emitted by large industrial units (power stations of gas or coal, incinerators, boilers...) and then inject it into geological structures so that it is stored there for very long periods of time*" [HA 10]. CCS technologies are considered as potentially available starting from 2020 [CHE 09]. It concerns expensive solutions developed for industrial sites. So far, the experimental nature of these technologies has been the subject of controversies, both at the level of their effectiveness as well as their economic viability: "*we already know that 20%, or a little more, of the total CO_2 can actually be recovered, and only the capture of fixed installations, that are more than a certain size, will be likely to meet the criteria of cost-effectiveness*" [FIN 10].

CCS technologies are a radical innovation, in the sense that they require both a change in the behaviour of the users and in companies' assets. Their implementation in energy systems goes through a strengthening of the technological systems of innovation [VAN 10a]. The control of risk associated with this type of innovation involves the development of close relations between industrial actors and scientific research. Thus the key actors in the development of this technology are multinational firms (Mitsubishi, Air Liquide, Exxon Mobil, Air Products and Chemicals, Kansai Electrical Power, Shell, Linde, Alstom and General Electric) and research organizations (US Department of Energy, Institut Français du Pétrole) [PIC 12].

2.4.2. *Producing energy differently from non-carbon resources*

The issue of an energy transition is also expressed in the choice of energy sources, particularly for electricity which sustains a significant portion of our economic activities. Facing the limits of fossil carbon resources (scarcity, cost, security of supply), the aim is to develop other sources of energy. Countries will make different choices of electric or other hybrid energy according to the resources available.

9 The concept of CCS concerns solutions of so-called geological storage, although other storage modes are theoretically possible such as ocean storage or storage through biological sequestration (biomass).

2.4.2.1. *Nuclear*

In France, nuclear energy was the first to appear as an alternative to fossil resources. If in the 1970s the fission of the atom represented energetic independence, more recently its contribution to the decarbonization of electricity is put forward. The major nuclearization programs began following the first oil crisis that highlighted the economic consequences of energy dependence [BEL 15]. From 45 reactors in service in 1965 in the world, we have gone to 365 (in 1985) and then to 438 in 2008. Despite the benefits attributed to this technology (independence of hydrocarbon prices, kWh inexpensive usage, no CO_2 emissions), this technology presents major industrial risks in the short-term (revealed by nuclear accidents) and poses questions regarding the long-term management of nuclear waste (nuclear power plants are only partially recyclable and radioactive waste has a high degree of environmental contamination). In France, currently 73% of electricity production comes from nuclear power stations. The deployment of nuclear power plants has enabled the French territory to be attractive by playing on the quality of the French energy system and on the level of the price of the kWh. Indeed, with the exception of Poland, Romania and Bulgaria, France is the country of the European Union where the price of the kWh of electricity is one of the lowest. Nuclear electricity is one of the least carbonated. However, the ageing of nuclear power plants poses issues and the difficulties inherent to the dismantling of these plants stresses the need to develop technologies by integrating the constraints on the whole of their lifecycle.

2.4.2.2. *Renewable energy*

In this context, new alternative technologies, including electric power, contributing to a decarbonization of energy, have been developed. They aim to promote the exploitation of renewable resources. Related to natural phenomena, renewable energies are inexhaustible primary energies. Whether wind, sun, water or wood, we are redeveloping old energy sources using technological advances to rehabilitate and increase their performance and reliability. So far, these technologies (hydropower, wind, biomass, solar, geothermal, marine) have different degrees of technological and commercial maturity. These energies, derived from natural phenomena, remain marginal on the scale of global consumption (from 14 to 20% according to sources) and French consumption[10], even if the objective is to increase their use to achieve the goals of the energy-climate plan[11]. According to the French energy roadmap, the share of renewables in final energy consumption is expected to reach 50% in 2020 (compared to 14.6% in 2010) and for the IEA, in 2050, 40% of the world's electricity will come from renewable energy.

Although the exploitation of renewable energy technologies are far from being completely neutral environmentally[12], their development occupies a place of choice in the directions being taken by the energy transition. From an economic point of view, the development of renewable energy questions their cost and their profitability in the short term. It is clear that their dissemination is highly linked to the actions of public policies. This is what Jacobsson and Lauber [JAC 06] demonstrated regarding the proliferation of wind and solar technologies in Germany. The dynamics of development of new energy technologies were the subject of various studies in the context of technological systems innovation: Jacobsson *et al.* [JAC 04b] and Dewald and Truffer [DEW 11, DEW 12], became interested in the photovoltaic, Bergek and Jacobsson [BER 03] and Markard and Petersen [MAR 09] have studied wind energy, Jacobsson [JAC 08], Negro *et al.* [NEG 07], Negro and Hekkert [NEG 08] and Wirth and Markard [WIR 11], worked on the biomass and Suurs and Hekkert [SUU 09b], Hellsmark [HEL 10] on biofuels.

However, from a technological point of view, one of the main difficulties of renewable energies (wind and solar PV) lies in their intermittent nature. Added to this is the fact that these energies are produced in a decentralized manner, unlike current systems, and impact the architecture of large energy systems. Accordingly, the large-scale use of renewable energy simultaneously raises the question of their connection to the current electricity grids, their

10 Key figures and statistical data: http://www.developpement-durable.gouv.fr/Chiffres-cles-et-donnees,29321.html; observations and statistics: http://www.statistiques.developpement-durable.gouv.fr/theme/energie-climat/966.html.

11 "*In 2014, from a total of 256.5 Mtoe of primary energy (in France), 9.4% originated from renewable sources. Renewable energy thus represents the fourth source of energy in the primary energy consumption, after the non-renewable primary (mainly nuclear) petroleum products and natural gas*" [COM 15c].

12 It should be recalled that the exploitation of renewable energies also has an impact on the environment: hydroelectric dams, extraction of rare earths, local environment pollution, recycling problems.

deployment within an efficient and intelligent electric network and the management of their intermittence, which goes hand in hand with the development of electric mass storage [PON 13]. Advances in the technologies of mass storage of electricity and in the increasing efficiency of electric networks management appear necessary and it is in this direction that European public funding programs of R&D are moving today [CAB 15].

2.4.2.3. *Mass storage*

Since it is derived from natural sources, renewable energy is an intermittent energy source with a gap between production and consumption. Indeed, solar or wind power allow the production of electricity at times when demand might be low. It is therefore important to be able to store this electrical energy. Currently there are chemical storage solutions (battery) but these solutions do not offer massive storage capacity. The solution generally used is a hydraulic one (dams), but its implementation is closely linked to geographical constraints that may not cover the areas of electricity production from renewable resources. The challenges of the development of these storage technologies as ground-breaking solutions seem to be emerging around hydrogen [BEC 15][13].

2.4.2.4. *Smart power grids*

The production of renewable energy goes hand in hand with a decentralized production of electricity. This type of production makes sense for isolated sites (particularly island territories) not easily integrated into the national network. However, the decentralized production of electricity shall eventually be reinserted into the national network. One of the challenges in the context of increasing the share of renewables in the mix is the management of the consequences of the introduction of renewable energies into the national electric grid.

Smart grids play a key role in the ability of the electric system to turn into a more diversified and more decentralized energy system (including micro-networks or local networks)[14]. A smart grid is an intelligent electrical network that coordinates the actions of individual users, producers, consumers, to supply these with a sufficient, secure and cost-effective

13 One of the technologies that has currently attracted particular attention is the use of hydrogen as a vehicle of energy storage. See, for example, the GRHYD project led by GDF-Suez: http://www.developpement-durable.gouv.fr/Transformer-en-hydrogene-l,30305.html.
14 http://www.developpement-durable.gouv.fr/IMG/pdf/E_D129.pdf.

electricity supply [EU 10][15]. Currently, in order to ensure the balance between the supply and demand of electricity which is not stored, a one-way logic of the producer towards the consumer prevails. However, the grids must adapt, become more "intelligent" by integrating new information and communication technologies, encourage two-way traffic in real-time information flows and enable a more efficient management of the electrical network in order to meet its challenges[16]. These smart grids combine energy and digital technologies and will eventually integrate renewable energy and electric devices to drive the consumption of households. They are therefore a critical element in the energy transition[17].

2.4.3. *Using energy differently: electromobility*

If the transport sector is to be heavily involved in the sustainable transition process, its role will be based on the concepts of mobility, especially individual mobility (passenger transport). The emergence of a service-based economy suggests the development of mobility rather than a need for vehicles. The demand for mobility would become substantial to the modernity of our societies as G. Amar indicated in his book *Homo Mobilis* [AMA 10]. Current modes of transport using internal combustion engines generate a considerable amount of environmental nuisance. It is therefore about organizing this evolution and to move from a "green transportation" logic (focused on "clean" engine technologies) to an ecological mobility. In this context, mobility takes on a more sociological meaning in the line of Urry's works [URR 05] or those of Kaufmann [KAU 08] and the question of the mobility transition, far from being summarized to that of a technological change, will treat the contents of travel and their societal modality. We are starting to observe societal innovations concerning the practices of mobility

15 The definition of smart grids is still in debate. If some studies limit the SG to a digital downstream solution of the meter and mostly aimed at residential customers, others have a broader vision. We could particularly refer to the white paper from Nexans (2010) or at the site of the Energy Regulatory Commission (http://www.smartgrids-cre.fr).

16 The challenges include the need to increase the production of electricity from renewable energies which is, intermittent and decentralized (which must be connected to the network and injected on the network as soon as it is produced), and the reduction of CO_2 emissions. The deployment of smart grids will contribute to the reduction of greenhouse gas emissions: reduction of 2.2 Gt of CO_2 per year by 2050 worldwide.

17 For examples of research projects or of experimental or commercial deployments of smart grids, see: http://www.developpement-durable.gouv.fr/IMG/pdf/E_D129.pdf and http://ses.jrc.ec.europa.eu/project-maps.

in concepts of carpooling[18] or car-sharing. A different vision of the relationship towards the automobile seems to create itself, which allows for the emergence of the business model of usage and not that of property [NIE 06].

However, many innovations aimed at enhancing the socio-technical system around the internal combustion engine by making it more effective, by feeding it biofuels [BRO 04] or by coupling it with an electrical battery. Simultaneously, radical innovations are also developed in the field of electromobility and more specifically with electric vehicles[19]. For the latter, several alternatives exist: vehicles using a battery[20] to produce electricity (BEV) and fuel cell vehicles (FCV). These older technologies have both experienced multiple hype and disappointment periods.

The transition to electromobility is marked by technological competition. Facing this technological competition [TRU 02, BRO 03, HEK 04], the transition of the current socio-technical system of transports towards a sustainable electromobilty can be analysed around two scenarios (one led by the need to regulate GHG emissions and the other led by the market) brought by these two technologies [VAN 10b]. These socio-technical scenarios take into account the endogenous nature of the dynamics of technological change and are shaped by the relationship between car manufacturers and their customers. Considering the long-term, Köhler *et al.* [KOH 09] concluded that FCV could dominate after 2030, unlike vehicles using biofuels and hybrid vehicles that may quickly prevail since they have the advantage of already being developed and marketed using the existing infrastructure.

---

18 Article L. 3132-1 of the ecological transition and green growth Act defines carpooling "*as the common use of a powered land vehicle by a driver and one or several passengers, conducted without remuneration, except cost-sharing, in the framework of a trip that the driver performs for his own account. Their relationship, to this end, can be conducted with remuneration and does not enter in the field of occupation defined in the article L. 1411-1*".

19 There is a diversity of electromobility solutions, for individual vehicles which mobilize different technologies: hybrid, rechargeable hybrid, electric (which itself is broken down into several categories). The 100% electric vehicles (without thermal power) are the latest technology for the electrification of the vehicle. When we talk about BEV we refer to vehicles that are 100% electric, not hybrids like Toyota's Prius model of a combustion engine and an electric battery.

20 Without going into detail, let us note that multiple battery technologies exist that differ in their chemical composition, the best-known being the lithium-ion battery.

The development of electromobility is led by users' acceptability of this type of mobility, especially for BEV [DUT 13, HAN 14, JAB 12, TSA 12]. A diversity of work relying on methodologies and different geographical areas allow the identification of the behavior of uses, or even their specificity from a socio-institutional point of view [ANT 14, CRA 12, DEL 12a, DUD 14, ENS 13, LIE 11]. Although it is premature to draw conclusions, the trend is that consumers generally have a positive image of these vehicles including their environmental performance (emission of locally polluting particles (NOx) and globally (CO_2))[21]. However, these studies stay biased in that they do not incorporate the issue of the environmental consequences of the development of transport infrastructures, which, by shaping the organization of cities, areas largely question societal issues of fair service accessibility and environmental issues in relation to the denaturalization of the soil and the reduction of biodiversity. In fact, the balancing of individual and collective mobility solutions is not paid much attention, nor is the integration of spatial organization issues of productive activities (sources of an increasingly daily professional mobility), their impact on the cost of the land and their consequences in terms of standards of living.

All historical manufacturers are today present in this market, and new actors have also made their entrance in recent years reactivating competition (Tesla, Bolloré, Google, to name only the best known) [WES 14]. We often observe that here as well innovation implies partnerships and joint ventures [VIT 12]. If the electrified vehicle market is growing, the growth is not as strong as it was initially thought to be. However, the works of GERPISA [GER 12a, GER 12b], and in particular those of Freyssenet [FRE 09], show that the hypotheses of a rapid transition to the "all-electric" car is possible. It notes three reasons. First of all, the battery technology can only improve and current offer allows covering a good part of the mobility needs. Then, the geopolitical and competitive arguments incite states to act as reducers of uncertainty and to guarantee risky investments. Finally, once the infrastructures are in place, the spreading of vehicles and the lowering of its production costs are achievable. On these emerging markets, we see all the difficulty that there may be in being "*first mover*" [BOR 14, MOU 15] and

21 An investigation by the Cetelem Observatory, carried out in 10 European countries, allowed us to know more about Europeans' perception of electric vehicles.

manufacturers are experimenting with new business models [KLE 11, BOH 14, MOU 15, JUL 12]. The emergence of these new business models is accompanied by a reflection relating to the evolution of smart grids [AND 09, BEE 12] and new energy technologies [BEL 12].

2.4.4. *Transition to a radically new energy system: the hydrogen economy*

Announcements are constantly being made about the imminent emergence of a hydrogen economy. This would involve a radical change in the economy's current dependence on fossil fuels.

2.4.4.1. *The particular case of hydrogen energy*

The development of a hydrogen[22] economy is regularly highlighted, with hydrogen being considered as the "fuel of the future" [KAL 13], able to substantially alter the ways in which we produce and consume energy. Presented as a solution to all problems [RIF 02], the provision of energy services based on hydrogen, just as much criticized as praised, generates cycles of hype and disappointment [ROM 04], which can be read in various public policies [BEL 14]. Hydrogen would not only provide energy to various sectors of the economy, whether transport, building or industry, but also provide storage solutions associated with intermittent renewable power technologies.

Although widely present in the universe[23], hydrogen does not exist in a pure gaseous state in nature. It occurs in combination with other atoms to then be separated according to various more or less expensive processes. It must therefore be produced from a primary energy source (oil, natural gas, methane, ethanol, etc.)[24]. This is why hydrogen is not a source of energy (or primary energy such as coal, oil, natural gas) but an energy vector (also called secondary energy source). Hydrogen can be produced from various

22 Industrial hydrogen has been long-standing; hydrogen discussed here is the hydrogen energy.

23 Hydrogen is the simplest element, the lightest, the most abundant of the universe which is 90% of the matter.

24 In this case, the development of a hydrogen economy would be sustainable; it would create additional pressure on fossil resources and increase GHG [HET 07].

energy sources and hydrogen-energy production can be done in different ways: extraction by steam reforming of primary energy, coal or biomass gasification, electrolysis of water. This last produce electricity process uses from conventional power plants – coal, gas, nuclear, etc. – or renewable energy – wind, solar, geothermal, hydraulic, etc. Accordingly, the hydrogen production process may not be neutral from an environmental point of view. In all the cases, environmental benefits of hydrogen relative to fossil fuels depend on technologies implemented to produce and use it. Beyond production technologies, there is also the question of the modalities of production. Two contrasting approaches exist, between a large-scale centralized production and smaller distributed installations on the territory in order to take advantage of local energy sources. The choices in this area influences its modes of transport (pipelines on short distances vs. cryogenic tanks for longer distances) and storage.

It is clear that these technological choices will depend on the ability of hydrogen to compete with fossil fuels or to build itself complementary to fossil. But beyond these techno-economic uncertainties, scientific and societal uncertainties remain. From a scientific point of view, our limited understanding of the hydrogen cycle generates uncertainty about the ecological consequences of the introduction of a hydrogen mass economy and the potential effects of the release of significant amounts of hydrogen in the atmosphere. Moreover, there is also a more societal problem, due to the idea of the danger[25] associated with this energy vector. Hydrogen is less flammable than petroleum products, but its low-density energy volume, its transport and storage in high or low temperatures involve other dangers. Due to these risks, the acceptance of this technology imply efforts of norms and regulation building. Scientists agree that the costs of a transition to a hydrogen economy are huge and that the transition would take several decades.

2.4.4.2. *The development of fuel cell technology*

Once produced, hydrogen is stored to be used to power a fuel cell that will convert it into electricity through an electrochemical reaction [BLU 09]. The principle of a fuel cell was revised simultaneously in 1838–1839, by two researchers, C.F. Schönbein from the University of Basel in

25 We often evoke the Hindenburg disaster (aircraft that crashed in May 1937 in the United States) but Addison Bain showed in 1997 that hydrogen was not at the origin of the accident, contrary to what has been often been claimed.

Switzerland and the English jurist, Sir W. R. Grove. It was only in the 1960s that F.T. Bacon made a first functional fuel cell (alkaline fuel cell) that would be subsequently used by NASA in its space programs. The use of fuel cells (FC) powered by a fuel rich in hydrogen and oxygen (Hy-FC) is part of the energy technology, which could promote the transition to a sustainable economy. Indeed, FC is an energy converter that produces electricity, heat from an electrochemical reaction. In order to produce enough energy FCs are traditionally connected in series within a stack. This stack is itself integrated in a more complex element composed of auxiliaries. Stacks and auxiliaries constitute the FC system. The FC can be fed from different fuels such as methanol, natural gas or hydrogen. It is regarded as a clean energy system because its operation emits no greenhouse gases or toxic gases due to the absence of combustion. Of course, this would involve hydrogen being produced cleanly from renewable energy and not from CO_2 emitting fossil resources.

Although it is not new, this technology is a disruptive one. In fact, it struggles to find a significant form of industrial deployment, beyond some niche markets such as forklifts where fuel cell technology is widely used. This technology can be mobilized to serve different application areas, including stationary, embedded, mobility applications or even energy storage[26]. Each of these areas allow to ensure one or more societal key functions (housing, mobility, communication) in which the Hy-FC competes with other (well established or emerging) technological solutions.

In recent years, a series of works have led on the applications of the FC. In the stationary field, fuel cells allow to produce heat and electricity using a cogeneration process [BRO 07, MAR 08a]. In the field of mobility, two major means of transport are studied. Buses appear as examples of an interesting demonstration for which the problem of the hydrogen supply is relatively simple to be solved by the establishment of a station near bus depots. At the same time, studies are also being done on the strategy of FC car manufacturers [MUS 12, KON 12]. Several joint ventures have emerged allowing builders to pool their resources and skills and receive public aid to

26 Hydrogen is not only a usable energy mass storage vector to produce energy, but it is also a technological option of energy mass storage (essential for supporting the development of renewable energies), alternative to currently used chemical storage.

support the development of this technology. The first marketing of private vehicles in Europe is very recent and still timely[27].

Transition studies explore the strategies of car manufacturers using fuel cell technology. R&D investments, which are dedicated to the development of this technology, especially since the beginning of the 1990s, cannot be understood with regard to the volume of the potential market, but more with expectations and visions of the future that build the actors in their strategic orientations [BUD 12]. Farla *et al.* [FAR 10] analyse the development of hydrogen and FC technology in the Netherlands between 1980 and 2007, mobilizing a Technological Innovation System (TIS) approach and show that TIS remained in the early stages of its construction over a long period. Despite the absence of short-term commercial prospects, technology remains research-driven and entrepreneur supported – elements that helped to cope with periods of disappointment and gradually develop this technological innovation system.

2.4.4.3. *Blockages*

The transition from a carbon system to a system involving hydrogen energy is the subject of a substantial number of works. Their analyses highlight the diversity of the sticking points, which basically refer to the questions about the ability of a hydrogen economy to emerge [SOV 10] and the ability of FC technologies to emerge beyond market niches.

Although technological barriers are shrinking, economic barriers still remain. The price of a service supplied from fuel cell using hydrogen, relative to competitive technological solutions in various fields of application, is still too expensive (excluding subsidies). A number of micro-economic papers showed that this technology, despite its interest, could not impose itself commercially before a few decades.

A second barrier is known as the chicken-and-egg dilemma[28], which questions processes of emergence between technology and dedicated infrastructure associated with it. In the case of Hy-FC technologies, it appears unwise to invest in a network of production and distribution of hydrogen when the uses remain poorly developed; similarly it appears difficult for consumers to buy a fuel cell vehicle without guarantees to be

27 Note the strategies of Mercedes and Toyota.
28 Found also in the case of the electric-battery vehicle [MEL 08, NAT 15].

able to feed it hydrogen. This is partly to deal with this dilemma that the industry actors focus on captive or professional fleets for which this issue is less blocking. At the same time, the prospective studies on the deployment of hydrogen infrastructure often show the existence of a snowball effect from a certain level of investment. It is to raise this threshold that local communities are mobilized on these projects.

When considering the systemic nature of innovation, other forms of blockages arise. This perspective is not about developing a technology (the FC) or an associated energy production (hydrogen). It is also necessary to invest in infrastructures of storage and distribution, to evolve or build the normative regulatory framework[29], to develop the services associated of maintenance, repairing and training. Finally, it is necessary to support uses to move from subsidized niche markets to commercial markets without public funding.

Institutional factors have some influence [BEN 10]. This is why it is particularly interesting to analyse the diversity of strategies implemented by the actors to accelerate the diffusion of this technology in different domains of application. While in France the technology barely grew and faced the turnaround of actors, in the United States the fuel cell is already well represented in some markets such as forklifts or back-up systems. In Japan, it is more stationary equipment that has developed. In Europe, the construction of a hydrogen corridor project is taking place and is designed to support the electromobility emergence, on which car manufacturers such as Mercedes-Daimler and Toyota are already present. Van den Hoed [VAN 07] stresses the need for convergence of several factors to encourage manufacturers to invest in this technology. It appears that a transition to a hydrogen economy requires heavy investments in infrastructure as much in production as in distribution and storage of hydrogen. These investments require strong governmental support and a guidance of national energy policies in favour of this energy source. In a complementary perspective, Konrad *et al.* [KON 12] show that in the absence of robust technological and commercial knowledge, the actors will rely on expectations [BAK 12, BOR 06], basically of collective nature, to make strategic decisions and coordinate their efforts.

29 For the automotive industry, see http://www.acea.be/statistics/tag/category/electric-and-alternative-vehicle-registrations.

2.5. Barriers of the energy transition

As we have just seen, the energy transition mobilizes a diversity of technical solutions with different levels of technological and commercial maturity. However, it is clear that barriers remain. Economists and sociologists have long stressed the key role of the socio-economic conditions in the adoption of innovations that is not governed by the only intrinsic quality of technology. How can we explain the difficulties of implementation of the energy transition? Why have previously presented technologies struggled to diffuse fully? The question arises from necessary conditions to allow to break from carbon lock-in [BEN 10, BEN 09, UNR 00], locks that make it difficult for the deployment of alternative solutions described previously. We will focus here on barriers that may be faced in implementing the energy transition.

2.5.1. *Market failures*

Barriers to the transition process can be partially associated with the definition of successful innovation. Consider the innovation (especially technological) according to a traditional acceptance in which the marketing of a new solution is the result of a succession of phases from the invention, creation, to innovation and distribution leads to research the points of blockage at each of the stages. The weakness of the invention and innovation will often be associated with a lack of R&D and will justify the public policy. More generally, it will be to seek sources of incentives to stimulate private investments. The weakness of dissemination will be considered as the consequence of either regulatory obstacles, or a market failure because of incomplete information and insufficient societal acceptance of the new technologies.

Indeed, in a standard economic approach, we can consider that what makes the energy transition delicate is the difficulty that the actors have to operate global choices in an uncertain and incomplete information situation. Prices are signals that will guide the decisions taken by the economic actors in the energy sector, even if it's imperfect signals. According to Rotillon, "[...] *economic decisions are made on the basis of the prices that determine the costs and/or the performance of possible alternatives, the change in the rules of the game return, at least in part, to change the rules of pricing.* [...]

the concept of externality characterizing precisely this situation where the prices don't reflect all the consequences of the decisions taken" [ROT 10, pp. 99–100]. The energy transition is based in part on the implementation of appropriate incentive devices, promoting the use of new technologies of energy, either in the form of price signals or regulatory constraints. In France, the tax approach is also mobilized to facilitate a change in the behaviour of households. The expected medium-term effects are both a reduction in energy consumption and a substitution between energies towards decarbonized energy. Thus the introduction of the carbon component in the taxation of fossil energies in 2014 induces a variation of taxes levied on energies based on their CO_2 content. According to the study by the Office of the Commissioner-General of Sustainable Development [COM 16], the introduction of the carbon component resulted on average in an increase of 3% of household energy bills in 2013 [COM 13].

Faced with the market's shortcomings, the investment in new energy technologies may be insufficient, especially with regard to environmental issues. The conventional terms of incentive to invest in these technologies will not suffice [MAR 06]. Beyond the imposition of carbon pricing to support the promotion of the new technologies of energy, the dissemination of low carbon technology goes through different solutions. Stern [STE 07] emphasized the need to combine three types of political actions to limit global warming by the diffusion of low carbon technologies: (1) charge a price for carbon – via the carbon tax, the market for the right to pollute, carbon contracts – in order to guide economic incentives towards low carbon solutions; (2) promote the innovation and diffusion of low carbon technologies by supporting R&D, to demonstration projects, by the creation of market niches; (3) move the institutional and non-market barriers to facilitate the dissemination of low carbon technologies. In fact, faced with the impossibility of a market and financial exclusive regulation to provide global solutions, new forms of world governance taking into account the different geographical scales of territorial construction are developing.

2.5.2. *Systemic failures*

In the framework of techno-ecologic transition processes, as we have shown in the previous chapter, the innovation considered is a system innovation: the question is not just how to develop and disseminate such or such new energy technology, but how to develop a whole technological system around low-carbon energy. Therefore, the nature of barriers to consider is different as pointed out by "transition studies".

According to the technological innovation system's approach, the emergence of a TIS is linked to the present actors; the interactions that they have in order to improve the technologies involved, change the institutional frameworks and develop new business models [KON 12]. Seven processes are key: the creation and dissemination of knowledge, the orientation of the directions of research, entrepreneurial experimentation, the formation of markets, the creation of a technological legitimacy, the mobilization of resources and the development of positive externalities [BER 08a, BER 08b, HEK 07, MUS 12]. Behind these concepts, it is a different approach to transition mechanisms that turns and that questions the ability of technological innovation systems emerging to challenge systems in place: "*over the years the analytical interest in TIS research shifted from general technological innovations contributing to the economic growth of countries to new (energy) technologies as nuclei for fundamental socio-technical transitions*" [TRU 12, p. 9].

The analysis of the process of transition in terms of niches, which, let us recall, consider that niches are incubators for emerging technological systems, also focuses on the influence of different factors that may impact the development of a new technology. By taking the case of the auxiliary power units powered by hydrogen, Agnolucci and McDowall [AGN 07] put forward three main factors: economies on an external scale, network effects and the behaviour and expectations of users.

Technologies are more or less objects of cyclical interest. The hype and disappointment cycles will eventually impact the ability of an innovation system to settle permanently on the basis of the commitment of actors and their technology support. In this context, the actors' expectations will guide the development of new technologies [ALK 12, BAK 12, BOR 06] and the strategies of these same actors [BUD 12]. They will also shape the trajectories of technological niches. Konrad *et al.* [KON 12] show that the craze for technology of fuel cells in the German-speaking countries is a key element in the formation of the innovation system. The constellation of actors involved will vary in relation to the collective interest in this technology.

In the implementation of the transition process, if the identification of a target seems to be a mobilizing factor, what matters is to detail the trajectories that lead to it [LOO 07]. Thus, the stalemate of the "Britain Green Vehicle" plan that aimed to make Britain a flagship of decarbonized mobility would be linked not only to the dissociation between the strategy of car manufacturers and public actors of the territory, but also to the lack of evidence as to the establishment of transformations [GER 12a, GER 12b]. Without a paradigmatic shift, transition processes will not take place. The energy transition cannot be reduced to technological substitution but rather it requires the mobilization of more complex recourses. Therefore, the current energy system and its governance must be rethought. It moved away from the energy supply oriented approach, with a centralized production and distribution of energy, an unquestioned belief on energetic abundance and a confidence in scientific and technological knowledge to provide solutions. In this context, the consumer was invited to participate in debates on energy and technological choices. This paradigmatic change involves other barriers, methodological, and questions even the form of debates, the ability of users to express in a non-partisan manner and to hear their needs, and that of experts to listen and actually put them in discussion. Indeed, how to define the orientation of long-term energy policy knowing that the future may not be the extension of the past, whereas it is necessary to break with the past model? Between the uses of quantitative model, the macroeconomic approach based on past trends and the use of purely qualitative approaches, the construction of a scenario whereby a convergence is delicate and questions the foresight approaches in the field of energy [DAV 14].

Beyond the more or less radical isolated innovations coming to feed a technology portfolio diversified by necessity, many authors defend the idea of the requirement to support system innovations around smart communities with a territorial unit (e.g. development at the local level of positive energy communities). Territories are central in the implementation of decarbonization and energy transition strategies, in a systemic perspective; the energy transition requires the association of technological, social, behavioural and institutional innovations [ANC 15]. These innovations, already developed today, are transforming energy systems deeply, including the relationships that are formed between actors, or even between countries. This is what allows highlighting a technological innovation systems approach.

3

Agro-ecological Transitions

The agriculture and food sector today faces many challenges [GAL 15] caused by consumer demand, due to the various food scandals of recent decades: public health issues that focus on the fight against increasing obesity, transparency requirements as to the sanitary quality of the products in relation to the exposure to pesticides and the need to guarantee the products' origin and their production, health/nutritional benefits and their environmental impacts; social issues built around issues of inequality and justice in relation to access to food or the social acceptability of certain technological innovations (GMOs, nanotechnologies); environmental issues such as air and groundwater pollution, emissions of greenhouse gases, protection of biodiversity, etc.; economic issues in terms of efficiency and competitiveness, and more generally in terms of coordination modes and dynamics determining the ability of stakeholders to address the challenges of sustainability and to propose innovations (in agricultural practices, industrial processes, in marketing channels, etc.).

These tensions echo the limits of the production model that was introduced in France after the Second World War. We recall that the postwar agricultural revolution led France to conduct a major intensification of its agriculture between 1950 and 1980 through development of mechanization and massive use of fertilizers and pesticides. The model, defined as "productivist", was accompanied by a specialization of farms made necessary by the increased technical level of farms aiming to provide greater labour productivity. To this technical specialization, a strong geographical specialization of agricultural systems is added (geographical separation between livestock and crop production, geographic concentration of manufacturing industries), which helped structure the agro-industrial sector and better control the quality, volumes and logistics of supplies. In each of the specialized systems, the number of cultured species has decreased and farmers have oriented their choice to the most profitable species, by using sanitary products to mitigate the negative effects of short rotations or monocultures [MEY 13a].

These developments take place in areas that have at one end consumers with high demands in terms of quality and product safety at a competitive price, and at the other end highly specialized farms that must be able to generate sufficient profitability. As we shall see in detail in this chapter, the agro-ecological transition based on the transition to a management system of a multifunctional and sustainable agriculture, enhancing agro-ecosystems, optimizing production and minimizing costs, thus appears as a possible answer. First, we review the definition of the term agro-ecology (section 3.1) and on the concrete forms that it takes at each of the main links in the chain: production, processing and consumption (section 3.2). If the issues around agro-ecology seem to unanimously attract attention, an agro-ecological transition is nonetheless faced with a series of obstacles that should be highlighted (section 3.3). We conclude our discussion by identifying potential levers for the agro-ecological transition in France, in particular by highlighting the role of public policy (section 3.4).

3.1. The notion of agro-ecology

Listed in the Loi d'avenir for agriculture in 2014 in France, agro-ecology is associated with a set of agricultural practices and models intended to design production systems providing at the same time high levels of productive, economic, environmental, health and social advantages [BOU 15]. This new discipline emerged in the first half of the 20th Century with the aim of linking agronomy and ecology.

3.1.1. *Towards an ecologically intensive agriculture...*

Today, agro-ecology is a concept that offers a global orientation, but whose definition is still vague and polysemic [STA 12]. It considers the crop and livestock systems as ecosystems where the natural environment is transformed by farming practices that, reciprocally, adapt to the environment. It seeks to understand the functioning of these systems controlled by humans, who use natural resources and rely on a web of interactions between living beings. This concept is close to that of ecologically intensive agriculture, characterizing an agriculture using sustainable natural processes and ecosystem functions, while maintaining a decent level of performance [BON 11].

While its main focus was centred around analysis "of agro-ecosystems" and sustainability (far exceeding the scale of the plot to be interested in all of the productive agro-ecosystem or "farming system") and on the production of knowledge and practices to make agriculture more sustainable, the field of study of agro-ecology has gradually extended to the entire food system, thereby integrating the organizational dimensions of industry and consumption [STA 12].

3.1.2. *... and a sustainable food regime*

In this context, agro-ecology can also be closer to the concept of sustainable food and sustainable food systems, also recently developed. Indeed, sustainable food regimes contribute to protecting and respecting biodiversity and ecosystems and are culturally acceptable, economically equitable, accessible, affordable, nutritionally safe and healthy and optimize natural and human resources according to the FAO (Food and Agriculture Organization of the United Nations)[1]. Sustainable food, defined as a set of practices, from production to consumption of foodstuffs, economically viable, socially sustainable and environmentally responsible [CHI 12], which are compatible, in turn, with the following criteria: food that is accessible to all, healthy and balanced, and meets human nutritional needs; a system that preserves the environment, climate, soil, water, and biodiversity; a diet that relies on the local, national, European and international level for sustainable agricultural production methods, ensuring a fair income for producers, and preserving the rural fabric and local development.

Let us now consider the implementation of agro-ecology, at each of the main links of the chain: production, processing and consumption.

3.2. The implementation of the agro-ecological transition

The agro-ecological transition is accompanied by a series of eco-innovations that unfold within sectors and territories. We will see now how they are implemented in the agricultural and food sectors, by focusing first

1 See http://www.fao.org/ag/humannutrition/biodiversity/fr/.

on the changes that affect agricultural production systems (section 3.2.1). We will then see what environmental innovations are developed in the agribusiness (section 3.2.2), and finally conclude with the noticeable changes in consumption patterns (section 3.2.3).

3.2.1. *Changing agricultural production systems*

3.2.1.1. *A low ecological modernization*

The strong negative impacts of agriculture on the environment, combined with the energy crisis and the slowdown of the production-agriculture model, lead to two forms of ecological modernization of agriculture [DUR 14]. The first is a “weak” form corresponding to the implementation of good practices to improve the efficiency of inputs, waste recycling or the “dissemination” of technology within precision agriculture. This form is positioned in the continuity of industrial agriculture in that it aims to reduce the main negative environmental impacts.

3.2.1.2. *A deep ecological modernization*

The second form called “deep”, also qualified as ecologically intensive [BON 11], is in opposition to the production model. It represents a paradigm shift [DOS 82] insofar as producers seek to replace conventional inputs (including chemicals) with the advantages provided by biodiversity in agro-ecosystems. This is to increase biodiversity at different spatial scales (agricultural plot in landscape) and time, so as to promote the interactions between plants, animals and microorganisms. The agro-ecological transition as we discuss here refers to this paradigmatic transformation.

It aims to promote sustainable production systems and develops through various new practices [LAS 15]. It may, for example, include crop rotation and integration (or, more accurately, reintegration) of legumes in farms, so as to reduce the use of pesticides (the question of introducing legumes will be the subject of a specific point in (section 3.2.1), the conversion of farms to organic farming, methanization[2] or agro-forestry[3].

---

2 Methanization is when the farmer uses the decomposition of organic matter to produce biogas and electricity while helping to reduce emissions of greenhouse gas farms and value the manure.

3 Agroforestry consists of promoting the hedges and trees at the edges of crops so as to make shade, and thereby reduce water requirements and produce natural fertilizers through the leaves that fall to the ground while enriching ecosystem biodiversity.

The implementation of ecologically intensive agriculture is complex because it is necessary to review all modes of conduct of farms and management of agrofood chains in a territory. Indeed, it requires disruptive innovations that are more complex to implement, and the development of new cropping systems, but also better coordination of stakeholders to develop the landscape structures or spatially distribute cultures. Furthermore, these changes occur in a situation of incompleteness of knowledge and of greater risk to farmers with the deployment of these new practices [DUR 14a]. We will return to these issues in section 3.3.

3.2.2. *Eco-innovations in agribusiness companies: a recent and moderate development*

In the context of climate change that is becoming increasingly tangible, governments have attempted to estimate the impact of various industries on global warming. In 2005, feeding practices (from production to processing of products and their purchase by the consumer) were the cause of 27% of emissions of greenhouse gases associated with overall household consumption [COM 11]. In addition, if the food initially evokes health issues in the minds of consumers, they claim to be increasingly interested in the environmental impact of the products they buy (BVA survey conducted in 2012 by ANIA). They place food products second after detergents in terms of the need for environmental information. The involvement of the food industry (IAA) in the development of environmental innovation approaches is particularly central in the context of a techno-ecological transition.

Faced with strong institutional and regulatory constraints [REN 00] and growing market pressure [BEI 05], agribusiness companies are increasingly incorporating more environmental innovations into their overall strategy, particularly in their innovation strategy. The explanatory factors of environmental innovations must be researched both on the side of internal factors (company size, membership to a group, R & D internal service, factors that play positively on the implementation of environmental innovations) and external factors (sector and market environment, particularly the geographic dimension of the market, technological environment, regulatory environment, geographical location of the company) [GAL 13].

However, it is clear that relatively little is known about the state of environmental innovations in the food industry. It may be noted that under the framework of innovation surveys of industrial enterprises, including the food industry, the issue of environmental innovation has only recently been introduced in 2006–2008 (CIS surveys (Community Innovation Survey))[4]. This shows the interest of policy in relation to environmental issues, and an increasing integration of environmental issues into the strategies of firms.

Let us first of all state that for agro-food enterprises, more than for industrial companies in general, the environmental concern is a reason for innovation. For 60% of French innovative food companies between 2006 and 2008, the realized innovation has a benefit for the environment compared to 50% for French companies as a whole. The larger the size of the company, the greater the consideration of the environmental factor increases: the figure is 80% for food companies with over 250 employees. This is particularly the case in the production process, since for 58% of considered food companies, the environmental benefit concerns the production process. Cost reduction is the primary objective of this strategy, with differences across industry sectors and size. In companies specializing in plant and animal oils and fats, waste, water and raw materials recycling and the reduction of soil, water or air pollution are the most frequently highlighted benefits (in six of 10 cases). Through the search for better performance, larger units target savings of materials, energy and CO_2. For 34% of companies, the environmental benefit of innovation is also at the stage of use of the product by the consumer. The recycling of the product after use is often the most advanced feature. It is concerned with 19% of product innovations or processes with an environmental benefit [LES 11].

Thus the analysis of the motivations for environmental innovation appears essentially economic. Consideration of the regulatory factor does not seem predominant. According to Gallaud *et al.* [GAL 12], the regulatory motivation to encourage companies to introduce environmental innovations in the food sector appears relatively weak. Contrary to the assumption of Porter and van der Linde [POR 95], the relationship between regulation and innovation cannot be considered automatic if the enforcement is not adequately controlled (for a discussion of this hypothesis see Box 1.3). Environmental regulation is generally constraining for the moment (Grenelle 2), although the Law on

4 These questions about environmental innovations are no longer included in the following CIS surveys.

Environmental Responsibility (LRE) leads to changes because it strengthens the responsibility of polluters. However, according to the CIS survey, the affirmation of the polluter pays principle does not lead to an increase in the use of preventive techniques and companies remain focused on the issue of reparation of environmental damage.

3.2.3. *Consumers committed to the environment*

Given the population growth, rising incomes and changing lifestyles that characterize the developed countries, production and food consumption exert increasing pressure on the environment [MIL 14]. Food crises in the late 1990s with the emergence of BSE ("mad cow disease"), the case of dioxin chickens and those of listeriosis and salmonellosis have helped to educate consumers about the issue of food quality. Concerned about environmental issues, food quality and health, some consumers have shifted their consumption behaviour towards more environmentally friendly food, for example organic agriculture, and the adoption of other distribution methods such as short food supply chains. We see that this trend has been growing for 10 years. Thus in 2015, organic food had a record rate of consumers and buyers. According to Agence Bio, nine in 10 French people (89%) consume organic produce, at least occasionally, compared to 54% in 2003. 75% consume these regularly, at least once a month, against 37% in 2003. In terms of purchasing, record levels are also achieved with more than eight out of 10 (82% vs. 43% in 2003) buying organic products, and 55% doing so regularly (vs. 33% in 2003)[5].

3.2.3.1. *The development of short food supply chains*

Short food supply chains illustrate a range of initiatives and social innovations that are particularly interesting. It was in Japan that these forms of short food supply chains were created in 1965 when women, worried about mercury contamination of food products, developed cooperation with farms committing to regularly buy their production. The "teikei" is the ancestor of short food supply chains [GAL 16]. From 1980, and particularly since 2000, that the model spread in the West, driven notably by the popularity of local partnerships solidarity between producers and consumers. They are called Gruppi di Acquisto Solidali (GAS: Italy), Association pour

5 http://www.agencebio.org/sites/default/files/upload/documents/5_Communication/Relations_Presse/dp_agencebio_def.pdf.

le maintien de l'agriculture paysanne (AMAP: France, Romania) or Community Supported Agriculture (CSA: United States, United Kingdom), and designate a direct agreement between farmers and buyers around a steady supply of seasonal products. They are an important development in France today even if short food supply chains purchases still account for only 6–7% of total consumer purchases [ADE 12].

The Ministry of Agriculture, Food and Fisheries defines short food supply chains in April 2009 as forms of selling mobilizing at most one intermediary between producers and consumers. The figures of the last agricultural census (2010) show that 21% of farms (hives, fruit, vegetables, vines, animal products) sold products in short food supply chains in 2010, especially as direct sales. The involvement in this marketing method depends on several factors. It is notably more common in small farms (excluding wine sector) and sectors of honey and vegetables (animal products are more demanding in terms of processing and preservation and are rarely sold in short food supply chains) [COM 13]. There are also large regional differences. The producing regions, where many operators are present on a production, traditionally commercialize in long marketing chain. Some examples of direct sales from producers to consumers can be found on the side of farm sales (basket-picking, farm market, etc.), collective sale (collective point of sale or collective basket), market sales, selling on tour, at home or with delivery relay points (workplace), by mail (Internet, etc.), sales that are organized in advance by AMAP (Association pour le maintien de l'agriculture paysanne) (Box 3.1), while catering (traditional, collective, etc.) or marketing by a dealer-retailer (butcher, convenience store, GMS, etc.) are examples of indirect sales with a single intermediary. The name "proximity short food supply chains" introduced a more limited distance criterion (generally less than 150 km) between the place of production and place of consumption [ADE 12].

The concept of "short food supply chains" includes a wide variety of production and marketing channels and these chains are supports of many economic and social issues [CHA 07a, CHI 12]. The social aspect of "proximity" is very much present in the food sector. The food, especially when it comes from the earth, carries emotional (taste), human (putting a face behind the product), and identity issues (development of the farming profession, development of a region). We know that we eat not only nutrients but also representations and imagination [FIS 90].

3.2.3.2. *The environmental benefit of the distribution mode by short food supply chains*

The issues of short food supply chains are also environmental, although the lack of studies and reliable indicators do not allow concluding categorically that the short food supply chains would systematically present a better environmental alternative than long supply chains. Hence, transportation has a lower environmental impact than the agricultural production stage itself [COM 13]. However, we note a number of positive points in short food supply chains: French productions are subject to national and European regulations that are among the most demanding; maintaining a peri-urban agriculture can help reduce urban expansion and preserve fertile soil; provide a security of supply; the impact on producers' practices through the direct link with the consumer who is more attentive to the inputs or the preservation of biodiversity. Also, according to ADEME [ADE 12], short food supply chains can reduce the packaging waste, and use less preservation processes because the time between production and consumption is reduced. Consumers buying in short food supply chains will increase their awareness of the environmental impacts of agricultural production, accept products more easily with a less perfect size but of good nutritional quality, and consume seasonal products (respecting the seasonality of products including "local" products is important because food produced in heated greenhouses consume more energy and emit more greenhouse gases). Finally, ADEME highlights the reduction of road transport of agricultural and food products, a quarter of transport in France concern these products.

Most farms that resort to short food supply chains are small and often involved in low-intensity logistics, close to organic agriculture or labelled organic. A survey of the Regional Federation of Organic Farmers from Brittany (FROB) estimates that 38% of the proportion of organic farmers sell in short food supply chains, representing only 7% of the farming population in Brittany [MAR 10]. Nationally 10% of producers selling in short food supply chains produce organic (only 2% in the long supply chains) and many of them have adopted practices relative to organic [ADE 12].

However, it is necessary to take into account the logistics for large volumes transported over long distances in an optimized manner (medium suitability of transport/delivery volume, delivery optimization, filling the truck) can have a greenhouse effect per transported ton lower than small amounts transported over short distances, but in poorly filled trucks and travelling empty [ADE 12]. Energy impacts related to the movement of consumers to buy the product is also a parameter to be considered. The advantage for the environment of organic farming is reduced if the car is used to make the purchase [RED 06].

> The AMAPs defend the logic of an agriculture that is "rural, socially equitable and environmentally healthy". They were developed in the late 1990s, a period marked by a succession of food crises and the importance of food safety. In fact, it was during this time that many consumers adopted purchasing modes, and consumption of different foods, as evidenced by the growth in consumption of organic products in the same period (+20% per year in the late 1990s) [LAM 08].
>
> The principle of AMAPs is simple. Each week the producers offer to a set of consumer members of the association a "basket" of products whose composition is based on production and its irregularities. Consumers agree to buy the basket throughout the duration of the season and pre-pay it, guaranteeing the producer the sale of his harvest and a steady income. The producer meanwhile guarantees consumers the freshest ingredients and production method (environmentally friendly). During these weekly distributions, each member fills his basket with a list of products developed by the producer, which depends on the harvest of the week[6]. The average quantity and price of this basket are fixed in a contract signed by each consumer with the producer in the early season, in general for half a year.
>
> AMAPs form the most visible of these alternative systems but there are also the shops of producers, farmer markets and various forms of short food supply chains defined as initiatives mobilizing one intermediary at most between producers and consumers. The purpose of these alternative systems is to oppose certain characteristics of the "dominant" system [DEV 10].

Box 3.1. *The AMAP: an example of an alternative agri-food network*

The transformations consisting of modifying existing practices, developing new farming systems and new forms of distribution upsetting the existing productive patterns of agri-food chains are faced with blockages that we intend to develop in the next point. Some of the levers, implemented in particular by public authorities, will be addressed in a second time.

6 Most AMAP operate like this in PACA (PACA or Provence-Alpes-Cote d'Azur is the region from which the majority of data have been collected in the work of Lamine and Perrot (2008)). The baskets' contents can also be chosen beforehand by the producer, and the producer can deliver the baskets without being present when consumers come to collect (the latter generally having a wider time range, giving more flexibility).

3.3. Obstacles and levers for the agro-ecological transition

One of the challenges of analyzing agro-ecological transitions lies in the identification of some of its blockages, and unlocking levers to promote greater sustainability of agri-food chains. The question that then arises is how to design and spread more agro-ecological practices. What is the impact of the emergence of new social, ethical and cultural values (fair trade, short food supply chains...) on value chains and their organization? What policy directions influence these developments? The objective of this section is to shed light on how these issues are addressed.

3.3.1. *Process of locking and unlocking*

In the tradition of evolutionary economics, technologies are part of the technological trajectories characterized by interweaving a multiplicity of actors involved in a given "technological regime": the interactions between scientific and technological innovations, and between economic factors and institutional actions generate effects which exclude alternative ways [DOS 82]. Thus, in the case of agriculture, the converging paths of different players have contributed to the emergence and strengthening of a production-technological regime [POS 96], based on a very high dependence on chemical inputs [MAG 13, MEY 13b]. This progressive locking of the production-technological regime explains the massive non-passage to integrated plant production (Box 3.2), alternative to chemical pesticides having been the subject of many experiments [COW 96]. Despite the negative externalities, both for the environment and health, generated by the use of pesticides, despite the emergence of technical problems (yield losses due to resistance to pesticides in the 1970s), their use continued to increase, farmers are "captive" to this chemical crop protection system and funding for research and development continues to focus on intensive farming systems and not on alternatives [WIL 01]. These factors explain that only a technology in place finds the necessary relays to its self-reinforcement and the phenomena of path dependency appears [DAV 85]. The "path dependency" translates as the stability of the dominant technologies and the fact that technologies with performance and similar functions, and perhaps superior long-term potential, can be put aside [LIE 95], due to effects of increasing returns. Arthur [ART 88] identifies five sources of increasing returns: learning through use (the more a technology is adopted, the more important the learning associated with its use becomes and the more efficient it becomes); network externalities (the more a technology is adopted, the more its value increases for the user because of the enlargement of the community of users); economies of scale due to the manufacturing in large series; the growth of information returns (the more a technology is known, the less risk aversion is a blocking factor for its use); and technological interrelationships (the structuring of the technical environment through the associated complementary technologies makes it more attractive).

The concept of integrated protection was developed 50 years ago with the objective of reducing the use of pesticides to minimize environmental impacts and costs while maximizing the economic results for the farmer.

The first European working group on integrated protection was established in 1959. It led to the creation of the regional group of the International Organization for Biological Control (IOBC), which established several working groups throughout Europe in the 1970s and 1980s to promote the implementation of integrated protection. In 2006, there were 20 in Western Europe and 16 in Eastern Europe. Agenda 21 [RIO 92] has integrated protection as an "optimal solution".

Integrated protection emphasizes the growth of healthy crops and consists of combining different methods (physical, chemical and biological) to fight against pests in order to reduce the use of pesticides. It allows the promotion of the natural mechanisms of protection and reducing risks to human health and the environment, ensuring the least possible disruption to agro-ecosystems (paragraph 6 of Article 3 of Directive 2009/128/CE of 21 October 2009 establishing a community framework for achieving a use of pesticides compatible with sustainable development). In France, it fit into the Ecophyto plan and is one of the illustrations approaches, "Produce Otherwise", initiated by the Ministry of Agriculture, Agro-Food and Forestry.

Box 3.2. *Integrated protection (http://agriculture.gouv.fr/protection-integree-pincipes-et-definitions)*

As we mentioned in previous chapters, particularly in the first chapter, Geels and Schot [GEE 07] proposed a theory of transition pathways according to which the transitions in a technological regime, result from interactions at three levels: niche innovations, changes in the socio-technical landscape that create pressure on the dominant regime, and niche innovations, which are windows of opportunity and have effects in terms of destabilizing the regime. These authors characterize different types of transition pathways depending in particular on the temporal coincidence between pressure on the regime and the emergence of niche innovations. Thus, when there is little pressure and a steady technological regime (few alternatives innovations), there will be a reproduction process, while at the opposite end when there is a lot of pressure and availability of innovations, we can see a deep reconfiguration plan. Between these two extremes, intermediate transition pathways are described like a reorientation of the regime "from within" (moderate change). In the agricultural and food sector, the question of whether or not there has been a profound reconfiguration of socio-technical system is asked.

Duru *et al.* [DUR 14a] propose to complete this analysis in terms of socio-technical systems with a conceptual framework that takes into account the collective management of natural resources, the conditions for its implementation at the local and farm level, and the governance of these management systems, which are, according to them, absent from this analysis. The concept of an "agro-ecologically regionalized system" highlights the crucial role of the players in the functioning of agricultural systems on a territory, on different types of hardware resources needed to run the system at all levels, of operating in the sector and the territory[7] and the cognitive resources that are intrinsic to the systems (knowledge, beliefs, procedures that actors use to define their individual and collective goals and strategies, and act upon). In this framework, the innovation process combines changes that are agronomic, socio-economic (change in the functioning of the actors system, cognitive resources, standards and informal agreements) and institutional (changing standards and laws). They require new modes of coordination and learning among actors in the sectors and regions, and adaptive management of farms, industries, and resources.

3.3.2. *Obstacles to the development of new agricultural practices*

Although a relative consensus has seemed to emerge on the need to change the agriculture model, many obstacles and block factors delay implementation of new practices and devices, at all levels of agrofood chain. In the case of

7 Formal standards, laws, regulations, and public policies are also included as hardware resources which have the purpose of framing, funding and promoting production methods and interactions between actors (for a complete presentation of agro-ecologically regionalized system, refer to the article of [DUR 14a]).

France, production mode changes constitute risks for farmers at a time when their income is rather low and many of them face significant problems [BON 11]. Uncertainties on both markets (high variation in prices, unstable international environment, competitive pressures) as well as at the level of the proposed techniques, often experimental, reinforce this prudence among actors. Furthermore, a significant proportion of the farming population is ageing, and they are few today. Young people wishing to start up find it difficult to do because of the cost of land and the necessary capital. In addition, social pressure can act unfavourably on the behaviour because it can be difficult for farmers to adopt practices that neighbours perceive as poor agricultural practices (presence of weeds, non-tillage, etc.) [GOU 10]. Finally, the marketing of products from ecologically intensive agriculture can be complex if the downstream stakeholders (processors, cooperatives, distributors) do not develop new promotion strategies themselves and do not organize themselves in another way to collect and sell products.

These different factors will come into play and contribute to slow down the development of new agricultural practices. This is particularly the case concerning the reintroduction of legumes in France, but also concerning the transition to a mode of ecological production.

3.3.2.1. *The difficult reintroduction of legumes*

The different agronomic and environmental benefits of legumes justify a greater importance of these species in the rotations of farms (Box 3.3). The nutritional properties of these species, both for the human consumption and animal feed, reinforce the interest in their use in our supply system [SCH 15]. Yet despite a number of public aids and more favourable periods for certain species, legume production continues to decline in France, and more widely in Europe, while in other countries like Canada, this production developed over the past two decades. The priority given to grain in the agricultural and commercial development since the 1950s, the intensification and specialization of agricultural production and French agro-business help to explain the marginalization of legumes into farming systems, at the upstream and downstream ends of the chain [MEY 13a]. By signing different agreements (Dillon round (1960–1962), Kennedy round (1964), Blair House (1992)), the European Community has allowed entry of soy without custom duties, used in animal feed.

Crop diversification (in the sense of diversification of species grown on a farm and/or territory) appears as a lever for action to reduce the use of external inputs in farming – pesticides, fertilizers, water – and the environmental impact associated with their overuse [CAV 09, MEY 13a, MEY 13b].

In this context, the cultivation of legumes (lentils, chickpeas, broad bean, field bean, lupine, alfalfa, soybean, etc.) has several interests. These plants do not require any nitrogen fertilizer for growth and restore this nutrient to the land for the following crops. Their presence in rotations helps consistently reduce the use of fertilizers, especially chemical ones. Moreover, these legumes are sources of protein for human and animal consumption. In the latter case in particular, they represent an interesting alternative to the very large import of soya cattle cake, mainly from Brazil and, to a lesser extent the United States, this strong dependence is of special concern in a period of growing demand from China on the matter.

We can see that despite their interest, availability of mineral nitrogen fertilizers and technical problems associated with mechanical harvesting of legumes have gradually declined in France since the 1960s. Also, although legumes were originally mostly cultivated for human food as protein sources, changing diets in favour of meat products have resulted in a decline of consumption of legumes from 7.3 kg/person/year in 1920 to 1.4 kg/person/year in 1985 [CAV 09]. Due to a lack of political and economic support for legumes, this production has become progressively confidential in France. However, despite low consumption, France still has a deficit in legumes (mainly lentils and beans) with a supply rate of only 27% on average from 2001 to 2008. Thus, in France, the situation is deteriorating over time: the surfaces for protein-rich crops have been declining since the early 2000s, according to the Ministry of Agriculture, Agro-Food and Forestry (plant proteins plan 2014–2020).

To enable the development of these cultures, the French government decided recently to implement the plant protein plan 2014–2020 for France. The stakes of this plan are various: to improve cooperation between the various stakeholders in production and processing, particularly animal feed manufacturers, specifically for the organic sector, and target protein independence for human food or animal feed. These objectives therefore require securing yields, and increasing them over time to develop the production of plant proteins; integrate in rotations crops that are rich in proteins, whether it refers to soybeans or alfalfa; secure opportunities, in the first instance animal feed but also other opportunities.

Box 3.3. *The revival of legumes: a sustainable solution for agricultural production?*

Despite the collective awareness of the issues that the model change presents, the associated cultures are barely present in conventional systems for field crops, particularly because they were disadvantaged in favour of more important investments in other cultures (both upstream and downstream of the value chains, especially in their industrial use): if different economic or environmental benefits lead to introducing more legumes in systems, this change will nevertheless not occur in production practices and the trend remains in the decline of surfaces.

Legumes have a low profitability for the farmer because their low yield does not sufficiently compensate the low need for fertilizers. On the other hand, agro-industrial sectors are displaying insufficient interest for their use and do not pay enough. The inclusion of legumes in crop production could contribute to the improvement of the economic performance of the farm, as soon as the knowledge is available and the systemic approach is taken into account [MAG 13, MEY 13a][8].

Another type of slowdown illustrates the systemic nature of agro-ecological transitions and relates to the logistical organization of cooperatives. This should be reviewed to the extent where it is not suitable for collection and marketing of this type of crops. As well as for the designation of organic farming production, and therefore the overall product quality[9] largely determining the remuneration of the farmer, harvests from practices labelled as "deviant" must be separated from other productions and more collection segmentation capability is thus necessary [MAG 13]. The logistics are therefore based on a specific organization and a quality management strategy. Magrini *et al.* [MAG 13] show that if the organization of cooperatives is, in principle, compatible with these new crops, we must also ask the question about the economic value of the products from these new practices–selection operations and multiplication of logistics flows to manage may engender significant costs. From the logistics point of view, sustainable logistics becomes one of the major issues of the Burgundian wine industry that sells 38.5% of its wines in the French market (outside the Burgundy region) and exports 47%. Already some foreign buyers (Great

8 Thus the introduction of a legume like peas generally increases the following crop yields and lowers overall costs (fertilizers and pesticides).

9 The quality classification criteria are multiple: moisture, test weight, protein content, impurities.

Britain, Canada, etc.) do not buy the wine if it does not respond to precise specifications in durability (labels, bottle weight, reduced use of transparent films pallets ...). Similarly, the development of sales in bottles by winemakers (excluding direct sales) leads to the multiplication of wine shipments in small quantities that can lead to negative environmental consequences with the increasing number of trucks on the road. Indeed, we know that sending larger amounts (in France or for export) may be less polluting than the distribution of fewer boxes more regularly [RIZ 10].

On the market side, obstacles of various kinds appear concerning the consumption of dried pulses and legumes products. One example is the lack of upstream competitiveness, which hampers the development of pulses in France; the strong dependence on the conditions of competition in globalized markets to marketing, as it is the case for soybeans; the small size of the industry and its lack of structure at all levels; a lack of a real industrial fabric supporting industrial transformation processes. The pulses are particularly marketed in "raw" form, such as whole seeds without any processing except the broken peas, which are husked and crushed. They are also consumed after canning with or without other ingredients (cooked lentils, preparations from white or red beans).

Ultimately, what demonstrates this first example is that the change of agricultural practices cannot be settled only by a regulation of the price system (privileged instrument of the common agricultural policy) but that structural changes affecting all the actors of the sector from the producer to the consumer are required [FAR 12, KLE 12]. The development of organic farming is another example of the difficulties of agro-ecological transition.

3.3.2.2. *The challenge of organic agriculture*

Organic agriculture is often presented as an alternative niche to the productivity model without calling into question the socio-technical system in place. However, one can observe a significant growth of organic farming in recent years. After a period of relative stagnation between 2002 and 2007, the sector was boosted from 2007 to 2011 with a strong increase in this sector [LAT 13]: +93% of the number of farms, +75% of the surfaces and +90% of the number of downstream operators. For all productions, the resulting share of organic farming continues to grow and organic certified surfaces went up to 1,118,190 ha in late 2014, an increase of 5.4% from 2013. Similarly, the conversion of organic agriculture producers over a 3-year process has been confirmed with 16% additional commitments in the first 5 months of 2015 compared to 2014[10].

On the distribution and consumption side, sales grew in all distribution channels. The highest growth rates were recorded in direct sales (+11.5%) and organic specialty stores (+11.3%) (http://www.agencebioorg/le-marche-de-bio-en-france). Thus more than half of the value of bio-products sales was made in the fresh food section. The market share of organic products varies depending on the sectors. In value, it is close to 20% with regard to eggs and 12% for milk. The market share of organic products is of the order of 7% for the 14 most consumed fruits and vegetables in France (excluding citrus fruits).

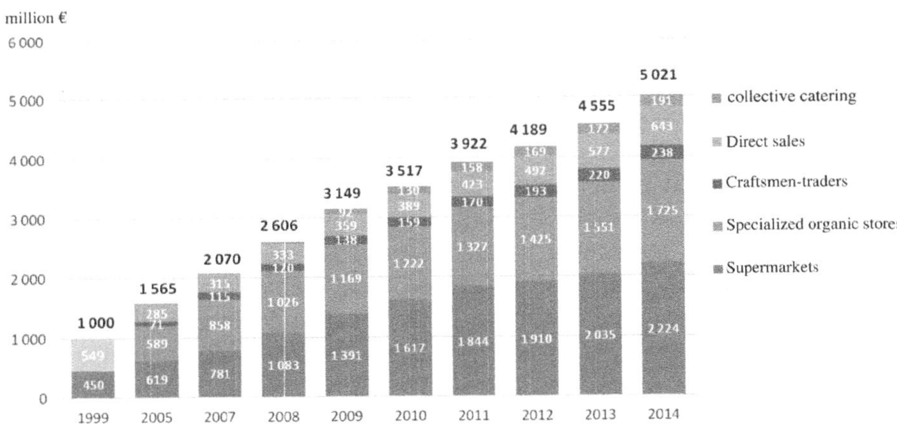

Figure 3.1. *Changes of sales by organic businesses in different distribution channels from 1999 to 2014 (source: Agence BIO/ANDI)*

However, despite this strong growth, organic agriculture still occupies a minor place in the agricultural landscape (4.9% of the Useful Agricultural Surface (UAS) of France at the end of 2015 and 5.7% on a European scale)

10 According to the Bio Agency (http://www.agencebio.org/la-bio-en-france).

and French food. Similarly, about a quarter of bio-products in France are imported. In the context of a promising market and strong political injunctions (Grenelle de l'Environnement de 2007), the conditions for facilitating the transition to this form of agriculture question remains.

The obstacles facing the development of organic agriculture (OA), raised by farmers at the agricultural production stage, include those relating to the lack of training and OA consulting, lower economic performances and the lack of solutions in place for plant protection and phyto-sanitary products of synthesis [LAM 10]. To these are added, separately according to the production in mind, the inter-annual variability of revenues and the crop protection for arable crops (cereals, oilseeds, protein crops, sugar beet, potatoes) in OA, high working hours in market gardening or in viticulture, or even the inadequate structuring of the sector and the lack of public support for dairy cattle breeding. Conversion to OA can, therefore, be seen as a decision that is "risky" financially [LEE 13]. An additional block to French OA and its competitiveness is the lack of harmony between European countries in contract documents and national regulation, according, to a study regarding strategy and foresight conducted at the request of the general Commissariat [GUY 13]. The study also mentions the obstacles at the stage of collection/processing (high collection costs, high prices of agricultural raw materials from the OA, irregularity of supply) and mentions the obstacles at the point of consumption (high prices, lack of information and points of OA product sales, confusion with some productions which are under official quality marks other than the OA).

3.3.3. *Blockages on the side of the agri-food sector*

Globally, the agri-food sector is a polluting sector [MAS 10a, MAS 10b], particularly in the treatment of wastewater and waste issues. For the moment, the measures adopted are therefore essentially remedial (in particular, the recycling of water) and not preventive. There is also another environmental issue, that of overuse of packaging where environmental innovations are scarce. Often, the agri-food manufacturers do not consider themselves responsible for this pollution, which is instead the responsibility of packaging manufacturers. Indeed, the lack of incentives to innovate in this field are reinforced because of the nature of collective good that the environment represents and, consequently, difficulties of private appropriation by corporations [GAS 03].

At the same time, investments aimed at reducing damage to the environment are generally low [MAS 09] even with the existence of regulation and the appearance of other sectors such as intermediate goods[11] that are more eco-innovating [GAL 13].

One of the explanatory factors for the relative weakness of eco-innovation in this sector lies in its structure, and particularly in the dominant presence of SMEs. In fact, these companies have more difficulties than larger firms in developing environmental innovations [UDA 08]. They consider that they are too small to have a negative impact on the environment [TEM 11]. In addition, they have more difficulties than large enterprises in benefiting from their innovations. The identified benefits of cost reduction, of quality improvement, and of opening new markets in particular were documented for large companies. The return on investment in environmental innovations takes a very long time [DEP 09], which requires the business to have a sufficiently large time horizon. This is not necessarily the case of SMEs whose survival and strategic vision are lower than those of large companies.

The IAA sector is a sector where production processes adapt progressively (slow technological change), because the sector's growth is gradual and adaptations are mainly incremental [DEP 09]. In this case, firms seek to reduce investment in new equipment and, on the contrary, to preserve as much as possible their production tool. This behaviour is often found in the most polluting sectors (chemistry, metallurgy, textile), which constitutes a serious barrier on the adoption of integrated preventive techniques.

From the point of view of the motivations underlying environmental innovation, literature relies heavily on the role of internal stakeholders such as employees, or of external stakeholders such as clients, suppliers, trade unions, professional organizations and non-governmental organizations (NGOs). In France, companies have little pressure from stakeholders for now [TEM 11]. Stakeholders would, once again, have more influence on large companies than on small ones. Lepoutre and Heene [LEP 06] point out

11 The industries of intermediate goods cover activities that produce goods most often intended to be reintroduced into other goods or which are destroyed by their use in producing other goods: mineral products; textile; wood and paper; chemicals, rubber and plastics; metallurgy and metal processing; electrical and electronic components.

that given the specificity of the small companies who for the most part consider mainly their immediate environment, relevant stakeholders who are most likely to have an impact are the customers and suppliers. The study of Quairel and Akhtar [QUA 07] shows that the power of customers in the context of eco-innovations and the power to pressure other actors in the supply chain are low at the moment. The power of shareholder pressure concerns SMEs very little and the role of consumers, trade unions, NGOs or local communities remains minor when it comes to encouraging businesses to innovate. As for the role of clients, the smaller companies did not cite it as a factor on the development of environmental innovations [GAL 12]. The main customers of SMEs are agri-business distribution groups. Finally, if the retail sector, a key client of agri-food businesses, played an important role in the dissemination of quality standards and norms, one wonders if it will have also a strong influence on better consideration given to the environment in companies.

3.4. The levers for agro-ecological transition: the role of public policies

The theory of socio-technical transitions [GEE 02, GEE 05b] led to the proposal to mobilize two major categories of levers in coordination: (1) the development of niches of innovation, places of realization for the process of learning and building new economic networks; these niches are intended to accommodate the construction and consolidation of channels for diversification; (2) the possibility within the standard socio-technical system of building opportunity windows, through which certain agro-food chains of diversification can grow and exceed the niche status. It is possible to imagine that these new sectors contribute to the evolution of the standard regime, and its transition.

Possible actions for one or another of these objectives [MEY 13b] relate to all of the actors in the sectors as well as public authorities. If the first of these actions is to encourage stakeholders to reduce inputs, the idea is also to increase crop diversification, for example via the common agricultural policy regulations, and promote channels of diversification through the public market. This impulse probably requires the development of criteria for the award of contracts. So, if on the consumer side the assumption is generally positive about the nutritional values of legumes, the disadvantages often seem prohibitive for the “average” consumer: lack of practicality, lack of culinary expertise or the image of an old-fashioned dish. A new effort to diversify products' images, similar to what has been done for cereals (breakfast, biscuits, energy bars, etc.) would make these products more attractive [SCH 15]. Legumes can thus play an important role in the nutrition transition, given the growing global demand for protein and the necessary balance between animal and vegetable proteins in the western diet.

Another measure is to promote the establishment and support of partnerships between actors of sectors, R & D, council, public research, selection and local authorities. These partnerships should be of a relatively long duration so as to build, at the local or regional level, diverse sectors advantageous to both territorial development and the environment (devices supported, for example, in the context of the second pillar of the Common Agricultural Policy). On the other hand, the implementation of the diversification of crops by farmers first requires in-depth studies on the part of research institutions and technical institutes on the choice of varieties, density seeding, fertilization and plant protection. Indeed, despite recent experimental studies carried out, in particular by the INRA, that show the performance of these crops from an agronomic point of view, it is today necessary to develop and offer to farmers technical pathways adapted to different objectives of production [MAG 13]. We must support genetic and technological innovation relating to diverse sectors that stimulate a reinvestment of public research in the selection of orphan species, and organize the partnership between public and private actors. Indeed, the difficulties that the farmers could face in their cultures (such as the irregularity of yields and health problems) are seen as particularly important, particularly when considering that significant improvements have been achieved for other species.

In order to support the development of species diversification, it is necessary to strengthen the system of specific terms (the first of which is official quality standards) for better marketing of the products of diversification crops to the final consumer, as is the case with the high environmental value (HEV) certification. This best value can also be achieved through the implementation of "agriculture-health" marker for products with high nutritional quality such as omega-3-rich animal products or dried vegetables. However, this differentiation may lead to transaction costs (collection, storage and traceability...) likely to reduce its economic interest, particularly when it comes to an offer of diversification crops spread out over a territory. The coordination of the actors along the chain is therefore essential for reducing these costs of transaction [MEY 13b].

One last proposal concerns the creation of an observatory of minority cultures in the territories and their place in the crop rotations and turns, so as to follow the process of diversification at work, and to be able to assess the agronomic, ecological and economic implications. If one of the issues of the agro-ecological transition is to give legumes and wider crop diversification a significant place, it is necessary for farmers to be equipped with a set of institutional approaches for the recognition of ecosystem services[12] that they can provide. There should also be provision of reference values adapted to the regional plan and accounting management tools adapted to the evaluation of the rotation.

If the role of all the actors of the agri-food sector is essential in the change of system, public policies contribute primarily, by creating a favourable, or even otherwise (when there are contradictions), a sustainable regime. Environmental regulation still concerns in a global, but reduced way the sector of agro-food industries (AFI). Grenelle 2 mainly targeted agricultural production: Ecophyto 2018 plan, environmental certification of farms, support for conversion to the organic production method, etc. On the other hand, AFIs were not referred to specific measures.

The agro-ecological project launched by the Minister of Agriculture in December 2012 was announced with an aim to make agro-ecology an asset for France. Its implementation was based on two main levers: the common agricultural policy (CAP) and the law for the future for agriculture, food and forest adopted by the National Assembly in September 2014. The ambition of the project was to develop a triple performance, economic, environmental and social, so that by 2025, a majority of farms are engaged in agro-ecology. This is to optimize the management of inputs and energy, diversify crops and strengthen the autonomy of animal production systems. Budget reorientation of the CAP should promote changes in practices through the greening of direct aid (first pillar of the CAP) and the mobilization of certain coupled aid, with priority given to agro-ecology in rural development (second pillar).

12 Pointing to the material benefits (goods directly consumable, self-purification of water, soil stabilization, pollination, etc.) or intangible (recreational or cultural activities for example) that the man collect from ecosystems ((http://ec.europa.eu/environment/pubs/pdf/ factsheets/Eco-systems%20goods%20and%20Services/Ecosystem_FR.pdf)

It is anticipated that incentives in the form of grants or loans are given to farms in the logic of "produce otherwise". Agro-ecology will be a priority of the national support measures financed by France AgriMer and the Office of Development of the agricultural economy overseas. Finally, the Loi d'Avenir for agriculture creates groups of economic and environmental interest (GEEI). These collective organizations will benefit from the CAP budget tools as well as tax advantages in the context of agro-ecological projects. To support this project, six action programmes have been strengthened or implemented, on plant protection products (Ecophyto plan), antibiotics (Ecoantibio plan), methanization, bee-keeping, vegetable proteins and organic agriculture (program national ambition bio 2017).

More recently, the French Government adopted at the Council of Ministers, on 4 February 2015, the national strategy of ecological transition towards sustainable development (NSETSD) 2015–2020. It replaced the national strategy of sustainable development (NSSD) 2010–2013, adopted in July 2010, and fixed a "new milestone" according to the Ministry of Ecology. Developing sustainable and resilient territories, engaging in a circular economy, strengthening education of sustainable development (EESD) and accompanying employment transitions, launching of new instruments for financing of the ecological transition or even reducing environmental, social and territorial inequalities are among the nine axes of the NSETSD shared across 34 priorities. Concerning agriculture, the draft dated March 20, 2014 shows that "*efforts should be strengthened in the efficiency of use of inputs that are chemical, energetic and of water, preservation of biodiversity, ecosystems and the functionality of soil carbon storage capacity. These constitute developments of agro-ecology, reconciling economic and environmental performance, and must be accompanied by the development of sustainable supply systems, contributing to a food supply that is safe and diverse, and to modes of production respectful of the environment*"[13].

A recent report referred to 30 projects for competitive agriculture that respects the environment and, as its name suggests, develops a series of measures designed to complement the agro-ecological transition in France and to provide agricultural stakeholders with research, training and the council regarding this new challenge (see Box 3.4). As we have seen, this transition also requires agreement on the part of the stakeholders, including those

13 http://www.consultations-publiques.developpement-durable.gouv.fr/IMG/pdf/Avant_ projet_ SNTEDD_20_mars_2014.pdf

involved in distribution and consumption. Thus it demands an integrative policy allowing actors of agriculture, agriculture-food, distributors and consumers to co-build a new model of production and consumption.

At the intersection of the agro-ecologic project for France and the national research strategy, ministers in charge of agriculture and research have designated four key figures to install the foundations of an "Agriculture – Innovation 2025" plan: Jean-Marc Bournigal, Managing Director of Cemagref, François Houllier, Managing Director of INRA, Philippe Lecouvey, President General Manager of ACTA, and Pierre Pringuet, President of AgroParis Tech. More than 300 actors were consulted in collective workshops and interviews: professionals of the agricultural world, researchers, actors from education, manufacturers, business and hubs of competitiveness, governments and regions, not to mention the downstream sector and civil society actors. The recommendations of the report produced in October 2015 are expressed through 30 projects subdivided into nearly 100 shares and aim at mobilizing the actors of research and innovation around major unifying issues. Three priorities have been identified, articulated around nine axes:

– Develop a system approach and make agriculture a contributor to the fight against climate change in: accompanying and stimulating the agro-ecological transition; developing research and innovation in the bio-economy[14];

– To enable the full development of the new technologies in agriculture by: continuing the digital revolution; accelerating the development of agricultural robotics; mobilizing the genetic lever and biotechnology; supporting the sector arising from the bio-control;

– Bringing together all actors in R&D, experimentation, and agricultural development in support of the competitiveness by: open innovation; assessing the multi-performance and innovative agricultural economics; supporting the actions of training.

Box 3.4. *30 projects for competitive and environmentally friendly farming*[15]

If all of the abovementioned projects proceed in the right direction, changes on the level of agriculture both impact and depend on changes at other levels of the "agricultural-food" system, a socio-technical system which includes farmers, councils, research, actors upstream and downstream

14 Defined here as the sustainable use of natural capital, the bioeconomy encompasses all production sectors processing and recycling of animal and plant biomass.

15 http://agriculture.gouv.fr/sites/minagri/files/rapport-agriculture-innovation2025.pdf.

of the sectors, public policy and regulatory bodies (regulation of phyto-sanitary measures, pollution, grains and the quality of the products in particular), consumers and civil society [LAM 12a]. However, as noted by Lamine and Chiffoleau [LAM 12b], food and agriculture are most often treated separately in public policy. A reconnection between these interrelated agricultural and food policies requires raising questions on these issues across territorial agri-food systems, since it requires the consideration and organizing of coordination on a territory of heterogeneous economic actors with conflicting of interests.

Ultimately, what technical and organizational innovations could allow the emergence and consolidation of innovative agro-food chains? What public action could promote the conceptualization of these innovations and the emergence of new sectors? What is the impact of these new fields on the ecological dynamics at the territorial level? Industrial ecology, circular economy and the analysis of territorial processes related to the agro-ecological transition of sectors contribute towards the renewal of analytical frameworks supporting the link between agro-food chains and territory.

A central obstacle to the adoption of innovative practices and transformation of the branches is the insufficient coordination of actors [GAL 15]. The unlocking of socio-technical systems to enable the development of changes such as the reduction of pesticides, for example, shows the importance of the coordination of actors throughout production and the importance of changes in practices of council but also in the organization of logistics or marketing. In this manner, collective strategies as new practices of sharing and pooling of equipment or products can be an interesting solution for structures that are too small to carry out innovative solutions. Faced with the issues of sustainable logistics, some wine sectors such as the comite champagne (CIVC) study these aspects closely by integrating into its objectives the consideration of the environmental impact of transport. It has thus implemented projects to increase exports by rail freight, optimize road freight and reduce transportation by air, and it has urged actions to reduce the weight of bottles [DES 07]. In the Burgundy branch, since the biggest companies are usually better armed than small ones to meet these challenges, some SMEs make use of technical and organizational innovations. Thus, small businesses that have experienced the distribution of their wines for export delegate some functions to logistics service providers with which they have a close relationship, forming lasting partnerships with relevant stakeholders, including clients, or developing innovations (such as the pooling of logistics or the use of carriers in batches). They improve their performance from a sustainable but also an economic point of view (cost reduction). If innovations exist and are implemented, their widespread use however, faces various obstacles. The most important ones are the increase in the number of marketers as well as the strong individualism in the Burgundy wine industry, which is mentioned very frequently and makes it difficult to implement collective strategies. In this context, cooperatives may make an exception, depending on a particular member's link and willingness to develop products "collectively" [TAN 15].

Indeed, according to Lamine [LAM 12a], it is particularly the very alternative and small-scale innovations that, in effectively echoing strong societal expectations, are developing with some success, such as basket schemes or collective point of sale systems. These measures are the sign of a real evolution of practices but are only a small part of the actors of this system and the population. It seems important in this particular framework not to oppose, as in theories of transitions, the "dominant" and "alternative" actors (or niches). Indeed, transition pathways are co-existing or competing with these actors and their types of initiatives, but they do not mix nor do hybrid systems emerge. However, Lamine shows the existence of hybrid systems, in particular through networks of organic and conventional producers or a co-operative store of organic products and its governance, aimed at involving a wide range of players in the sector. Finally, it stresses the combination of "top down" (from institutional impetus) and "bottom up" dynamics (from the private or collective initiatives of professionals or civil society). We are therefore witnessing a proliferation of initiatives, actions and projects, not necessarily coordinated, seeking to reform the links between city and agriculture, between power and territories [DEL 12]. The diversity of initiatives, private and public actions, and of their scales, on the one hand, formulate the question about their coordination, and, on the other hand, that of the emergence of new models for governance of agricultural and agri-food sectors.

It seems therefore essential today to show interest in the process of transition itself, which is less evolved rather than the states of balance referred to or achieved [LAM 09b]. Cases of transition to organic farming and integrated protection analysed by Lamine *et al.* [LAM 09b] thus show the importance of learning and collective dynamics concerning relevant sizes (productions, soil, rotate, phytosanitary practices, labor organization, marketing, learning, social networks) by these changes. As noted by Magrini *et al.* [MAG 13, p. 28], we can consider that "the accumulation of incremental innovations – for example, within a cooperative as a result of the acceptance of new practices among farmers – is likely to generate a combination of knowledge and specific resources constituting the basis for a transition, which, by hybridization, could lead to a switching of the entire agricultural system to a new technological paradigm". We can then speak of a gradual transition to a new system by making use of new knowledge acquired by the actors, which can overcome obstacles related to the dependence of a historically constructed path (path dependency).

Conclusion

This book has focussed on the issues concerning energy and agro-ecological transitions. These two areas are today confronted with the limits of current production systems whose environmental impacts are no longer sustainable. They have been the subject of a numerous of studies since 2000. A variety of research trends are also focussed on this issue. The transition can be approached through its political objective and the construction of scenarios of society: it is then referred to as sustainable or environmentally friendly. It can be seen in the context of its areas of application or major societal functions. It then becomes transitional agro-ecological, energy and even mobility-based. Finally it can be analyzed through a reading of its constituent elements: innovations, technological systems and socio-technical systems.

Our approach to these transitions led us to focus on their techno-ecological character, based on the approach of "transition studies". In doing so, three key ideas emerge as to what characterizes the techno-ecological transitions. First, techno-ecological transitions cannot be summarized as a process of adaptation of existing socio-technical systems, but it builds on innovation systems in which eco-technological and non-technological innovations create real environmental value [BOO 10]. This environmental value led economic actors to change their business model towards higher circularity and functionality. Second, in viewing technology systems as social constructions which address a number of functions, techno-ecological transitions require the development of new economic practices. From this perspective, eco-innovations may not be isolated or decontextualized. It is therefore important to consider the process of societal integration of these new technologies beyond a simple questioning of their acceptability, and to assess how this integration requires a change in scope, representation and practice for all stakeholders involved. Third, this leads us to remember that no matter how interesting and powerful the new technological solutions could be, a global evolution of behavior is necessary in both industrialized and emerging countries. Any technological change is accompanied by changes that are societal, organizational and institutional and this is especially true when considering structural changes.

When faced with challenges of sustainability, the ability of actors to innovate is central. As we have seen, transition mobilizes a variety of technical solutions with different degrees of technological and commercial maturity. However, it is clear that barriers remain. We have seen that the difficulties of implementation of the techno-ecologic transition originate in particular from progressive locking of the current technical and organizational systems. In the tradition of evolutionary economics, technologies fit into technological trajectories, and economic and institutional factors create systems which generate effects of exclusion of alternative pathways, despite the negative externalities generated. In the field of production and consumption of energy, obstacles associated with the use of carbon sources are numerous, when chemical inputs and the productivity model control the agricultural sector. Self-reinforcing, path dependency processes involve the stability of leading technological regimes. Developed within niches, new technologies are struggling to replace them. Transformations will be achieved in the coming together of emerging eco-innovations and environmental pressure, which will destabilize the existing regime. The need for systematic changes requires the intervention of public actors at all levels, locally and globally.

We have seen that regardless of the field in question, the techno-ecologic transition questions deeply the very foundations of our societies. The question legitimately arises of the ability of resilience of the territories and companies [HOP 10]. According to some, in facing multiple uncertainties related to climate change and its consequences, contemporary societies are likely to prioritize managing adaptation before thinking of their structural transformation. It would then follow to evolve organizations, spatial locations and activities of individuals, the techniques used in order to limit the negative direct effects of climate change, involving, for example, the adoption of more resistant plant varieties, the displacement of populations and productive sites of flood-prone areas, adaptation of grids to changes in energy consumption [HAL 10]. In reality, there is no contradiction here. Transitions should not be thought of only as a future state to achieve but as a constant process of change.

They carry with them the idea of an ongoing adaptation of our societies and productive processes in the context of uncertainties on the future evolutions of the climate, their impacts, the resilience of actors, alongside questions about technological evolutions. Therefore, it is not only about thinking of large-scale impact transformations, structural as we have dealt with here, but to question the process of movement of human communities, territories and to support their ability to develop new solutions, to imagine, and to build tomorrow's society. From this perspective, the territories are at the forefront, and in particular sustainable cities [EME 07], where an urban and periurban agriculture is settling. This change is accompanied by the need for a local expression, territorialized by the transformation of our societies, which is reflected by new forms of governance.

Bibliography

[ABE 85] ABERNATHY W.J., CLARK K.B., "Innovation: mapping the winds of creative destruction", *Research Policy*, vol. 14, pp. 3–22, 1985.

[ADE 12] ADEME, Les circuits courts alimentaires de proximité, Les avis de l'ADEME, available at: http://www.ademe.fr/sites/default/files/assets/documents/avis_ademe_circuits_courts_alimentaies_proximite_avril2012.pdf, April, 2012.

[AGH 09] AGHION P., HEMOUS D., VEUGELERS R., "Quelles politiques pour encourager l'innovation verte?", *Regards croisés sur l'économie*, no. 6, pp. 165–174, 2009.

[AGN 07] AGNOLUCCI P., MCDOWALL W., "The technological change in niches: Auxiliary Power Units and the hydrogen economy", *Technological Forecasting and Social Change*, vol. 74, pp. 1394–1410, 2007.

[AHM 08] AHMAN M., NILSSON L.J., "Path dependency and the future of advanced vehicles and biofuels", *Utilities Policy*, vol. 16, no. 2, pp. 80–89, 2008.

[AIT 10] AIT-DAOUD S., LAQUECHE J., BOURDON I. *et al.*, "Ecologie et technologie de l'information et de la communication: une étude exploratoire sur les éco-TIC", *Management et Avenir*, vol. 9, no. 99, pp. 307–325, 2010.

[ALB 09] ALBRECHT J., *Transition énergétique: plus vite vers un système plus vert?*, Itinera Institute, Ghent, 2009.

[ALK 12] ALKEMADE F., SUURS R.A.A., "Patterns of expectations for emerging sustainable technologies", *Technological Forecasting and Social Change*, vol. 79, no. 3, pp. 448–456, March 2012.

[AMA 10] AMAR G., *L'homo mobilis. Le nouvel âge de la mobilité – Eloge de la reliance*, FYP Editions, Paris, 2010.

[AMB 09] AMBEC S., LANOIE P., "Performance environnementale et économique de l'entreprise", *Économie et Prévision*, vols. 4/5, nos. 190–191, pp. 71–94, 2009.

[ANC 15] ANCRE (Alliance Nationale pour la Coordination de la Recherche en Energie), Decarbonization wedges, Report, pp. 56, November 2015.

[AND 05] ANDERSEN M.S., "Decoupling environmental pressures and economic growth", *Public Policy Research*, vol. 12, no. 2, pp. 79–84, 2005.

[AND 08] ANDERSEN M.M., "Eco-Innovations: towards a taxonomy and a theory", *DRUID Conference Entrepreneurship and Innovation*, Copenhagen, Denmark, pp. 16, June 17–20, 2008.

[AND 09] ANDERSEN P.H., MATHEWS J.A., RASK M., "Integrating private transport into renewable energy policy: the strategy of creating intelligent recharging grids for electric vehicles", *Energy Policy*, vol. 7, pp. 2481–2486, 2009.

[ANT 14] ANTIKAINEN M., "Electromobility and transport in Finland", NVF Meeting, 22 April 2014.

[ART 88] ARTHUR B., "Competing technologies: an overview", in DOSI G. *et al.* (eds), *Technical Change and Economic Theory*, Pinter Publishers, London and New York, pp. 590–607, 1988.

[BAK 12] BAKKER S., VAN LENTE H., MEEUS M.T., "Credible expectations – the US Department of Energy's Hydrogen Program as enactor and selector of hydrogen technologies", *Technological Forecasting and Social Change*, vol. 79, no. 6, pp. 1059–1071, 2012.

[BAT 03] BATHELT H., GLUCKER J., "Toward a relational economic geography", *Journal of Economic Geography*, vol. 3, pp. 117–144, 2003.

[BAT 06] BATAILLE C., BIRRAUX C., "Les nouvelles technologies de l'énergie NTE et la séquestration du CO_2", Report, OPECST, March 2006.

[BEC 15] BECHERIF M., RAMADAN H.S., CABARET K. *et al.*, "Hydrogen energy storage: new techno-economic emergence solution analysis", *Energy Procedia*, vol. 74, pp. 371–380, 2015.

[BEE 12] BEER S., GOMEZ T., DALLINGER D. *et al.*, "An economic analysis of used electric vehicles batteries integrating into commercial building microgrids", *IEEE Transactions on Smart Grids*, vol. 3, no. 1, pp. 517–525, 2012.

[BEI 05] BEISE M., RENNINGS K., "Lead markets and regulation: a framework for analyzing the international diffusion of environmental innovations", *Ecological Economics*, vol. 52, pp. 5–17, 2005.

[BEL 12] BELLEKOM S., BENDERS R., PELGRÖM S. *et al.*, "Electric cars and wind energy: two problems, one solution? A study to combine wind energy and electric cars in 2020 in the Netherlands", *Energy*, vol. 45, pp. 859–866, 2012.

[BEL 14] BELOT R., PICARD F., "The three ages of fuel cell research in France: a socio-technical system's resistance to change", *Fuel Cells Journal*, vol. 14, no. 3, pp. 525–532, 2014.

[BEL 15] BELOT R., *L'atome et la France. Aux origines de la technoscience française*, Editions Odile Jacob, 2015.

[BEN 09] BENTO N., ANGELIER J.P., "La transition vers l'hydrogène est-elle bloquée par un verrouillage technologique au profit des énergies fossiles?", *Cahier de Recherche LEPII*, no. 27, pp. 21, November 2009.

[BEN 10] BENTO N., "Is carbon lock-in blocking investments in the hydrogen economy? A survey of actors' strategies", *Energy Policy*, vol. 38, pp. 7189–7199, 2010.

[BER 02a] BERGEK A., Shaping and exploiting technological opportunities: the case of renewable energy technology in Sweden, PhD Thesis, Department of Industrial Dynamics, Chalmers University of Technology, Gothenburg, 2002.

[BER 02b] BERKHOUT F., "Technological regimes, path dependency and the environment", *Global Environmental Change*, vol. 12, pp. 1–4, 2002.

[BER 03] BERGEK A., JACOBSSON S., "The emergence of a growth industry: a comparative analysis of the German, Dutch and Swedish wind turbine industries", *Change, Transformation and Development,* Physica-Verlag HD, pp. 197–227, 2003.

[BER 07a] BERGEK A., JACOBSSON S., HEKKERT M., "Functions in innovation systems: a framework for analysing energy system dynamics and identifying goals for system building activities by entrepreneurs and policy makers", in FOXON T., KÖHLER J., OUGHTON C. (eds), *Innovations for a Low Carbon Economy: Economic, Institutional and Management Approaches*, Edward Elgar, Cheltenham, 2007.

[BER 07b] BERGEK A., JACOBSSON S., HEKKERT M. *et al.*, "Functionality of innovation systems as a rationale for, and guide to innovation policy", in SMITS R., KUHLMANN S., SHAPIRA P. (eds), *Innovation Policy, Theory and Practice. An International Handbook*, Elgar Publishers, 2007.

[BER 08a] BERGEK A., JACOBSSON S., CARLSSON B. *et al.*, "Analyzing the functional dynamics of technological innovation systems: a scheme of analysis", *Research Policy*, vol. 37, pp. 407–429, 2008.

[BER 08b] BERGEK A., JACOBSSON S., SANDEN B., "'Legitimation' and 'development of positive externalities': two key processes in the formation phase of technological innovation systems", *Technology Analysis and Strategic Management*, vol. 20, no. 5, pp. 575–592, 2008.

[BIJ 87] BIJKER W.E., HUGUES T.P., PINCH T.J., *The Social Construction of Technological Systems: New Directions in the Sociology and History of Technology*, MIT Press, 1987.

[BLU 09] BLUNIER B., MIRAOUI A., *20 questions sur la pile à combustible*, Technip, 2009.

[BOH 14] BOHNSACK R., PINKSEB J., KOLK A., "Business models for sustainable technologies: exploring business model evolution in the case of electrical vehicles", *Research Policy*, vol. 43, no. 2, pp. 284–300, 2014.

[BON 03] BONTEMS P., ROTILLON G., *L'économie de l'environment*, La Découverte, Paris, 2003.

[BON 11] BONNY S., "L'agriculture écologiquement intensive: nature et défis", *Cahiers Agricultures*, vol. 20, no. 6, pp. 451–462, 2011.

[BOO 10] BOONS F., *Creating Ecological Value. An Evolutionary Approach to Business Strategies and the Natural Environment*, Edward Elgar, 2010.

[BOR 06] BORUP M., BROWN N., KONRAD K. *et al.*, "The sociology of expectations in science and technology", *Technology Analysis and Strategic Management*, vol. 18, pp. 285–298, 2006.

[BOR 14] BORUP M., "Electro mobility case study for Denmark. Case study: better place: an effort of creating new actor roles and infrastructure for electric car mobility", TOP-NEST Project report, April 2014.

[BOU 14] BOUTILLIER S., LAPERCHE B., PICARD F., "Le développement des systèmes produits-services dans les entreprises: une étape vers l'économie de la fonctionnalité?", *Economies et sociétés*, vol. 48, no. 4, 2014.

[BOU 15] BOURNIGAL J.M, HOULLIER F., LECOUVEY P. *et al.*, "Agriculture/ Innovation2025 – 30 projets pour une agriculture compétitive et respectueuse de l'environnement", Report, available at: http://agriculture.gouv.fr/sites/minagri/ files/rapport-agriculture-innovation2025.pdf, October 2015.

[BRO 03] BROWN H., VERGRAGT P., GREEN K. *et al.*, "Learning for sustainability transition through bounded socio-technical experiments in personal mobility", *Technology Analysis and Strategic Management*, vol. 15, pp. 291–315, 2003.

[BRO 07] BROWN J.E., HENDRY C.N., HARBORNE P., "An emerging market in fuel cells? Residential combined heat and power in four countries", *Energy Policy*, vol. 35, pp. 2173–2186, 2007.

[BRU 87] BRUNDTLAND G.H., "*Notre Avenir à Tous*", report by the commission mondial sur l'Environnement et le Développement, Les Editions du Fleuve, Montréal, 1987.

[BUD 12] BUDDE B., ALKEMADE F., WEBER K.M., "Expectations as a key to understanding actor strategies in the field of fuel cell and hydrogen vehicles", *Technological Forecasting and Social Change*, vol. 79, no. 6, pp. 1072–1083, 2012.

[CAB 15] CABARET K., PICARD F., "Politique européenne de l'énergie et transition vers un système énergétique décarboné fondée sur les Smart Grids", *Revue Innovations*, vol. 1, no. 46, pp. 33–69, 2015.

[CAL 86] CALLON M., "The sociology of an actor-network: the case of the electric vehicle", in CALLON M., LAW J., RIP A. (eds), *Mapping the Dynamics of Science and Technology – Sociology of Science in the Real World*, Sheridan House, Basingstoke, 1986.

[CAR 91] CARLSSON B., STANKIEWICZ R., "On the nature, function, and composition of technological systems", *Journal of Evolutionary Economics*, vol. 1, no. 2, pp. 93–118, 1991.

[CAR 02] CARLSSON B., JACOBSSON S., HOLMÉN M. *et al.*, "Innovation systems: analytical and methodological issues", *Research Policy*, vol. 31, no. 2, pp. 233–245, 2002.

[CAV 09] CAVAILLES E., *La relance des légumineuses dans le cadre d'un plan protéine: quels bénéfices environnementaux?*, Etudes & Documents, no. 15, Commissariat Général au Développement Durable, MEEDDM, available at: www.developpement-durable.gouv.fr, p. 44, 2009.

[CHA 07a] CHAFFOTTE L., CHIFFOLEAU Y., "Vente directe et circuits courts: évaluations, définitions et typologie", *Les cahiers de l'Observatoire CROC*, INRA, Montpellier, vol. 1, no. 8, 2007.

[CHA 07b] CHARTER M., CLARK T., *Sustainable Innovation: Key conclusions from sustainable innovation conferences 2003–2006*, The Centre for Sustainable Design, University College for the Creative Arts, pp. 48, 2007.

[CHE 09] CHEVALLIER B., "Conditions de captage et de stockage du CO_2", *Pétrole & Gaz Informations*, no. 1799, pp. 44–46, 2009.

[CHE 13] CHEVALIER J.M., CRUCIANI M., GEOFFRON P., *Transition énergétique – les vrais choix*, Editions Odile Jacob, Paris, 2013.

[CHI 12] CHIFFOLEAU Y., PREVOST B., "Les circuits courts, des innovations sociales pour une alimentation durable dans les territoires", no. 224, Norois, Environnement, aménagement, société, pp. 7–20, 2012.

[COE 10] COENEN L., DIAZ LOPEZ F.J., "Comparing systems approaches to innovation and technological change for sustainable and competitive economies", *Journal of Cleaner Production*, vol. 18, no. 12, pp. 1149–1160, August 2010.

[COE 12a] COENEN L., BENNEWORTH P., TRUFFER B., "Towards a spatial perspective on sustainability transitions", *Research Policy*, vol. 41, pp. 968–979, 2012.

[COE 12b] COENEN L., TRUFFER B., "Places and spaces of sustainability transitions: geographical contribution to an emerging research and policy field", *European Planning Studies*, vol. 20, no. 3, pp. 367–374, 2012.

[COM 13] COMMISSARIAT GÉNÉRAL AU DÉVELOPPEMENT DURABLE, "Consommer local, les avantages ne sont pas toujours ceux que l'on croit", *Le point sur*, no. 158, p. 4, March 2013.

[COM 15a] COMMISSARIAT GÉNÉRAL AU DÉVELOPPEMENT DURABLE, "Les émissions de CO_2 dues à la combustion d'énergie dans le monde en 2012", *Chiffres & statistiques*, no. 595, January 2015.

[COM 15b] COMMISSARIAT GÉNÉRAL AU DÉVELOPPEMENT DURABLE, "Transition(s) vers une économie écologique. Le programme de prospective", *Etudes et documents*, no. 129, p. 52, July 2015.

[COM 15c] COMMISSARIAT GÉNÉRAL AU DÉVELOPPEMENT DURABLE, *Chiffres clés des énergies renouvelables – Edition 2015*, Repères, December 2015.

[COM 16] COMMISSARIAT GÉNÉRAL AU DÉVELOPPEMENT DURABLE, "L'impact, pour les ménages, d'une composante carbone dans le prix des énergies fossiles", *Le point sur*, no. 225, p. 4, March 2016.

[COW 96] COWAN R., GUNBY P., "Sprayed to death: path dependence, lock-in and pest control strategies", *The Economic Journal*, vol. 106, pp. 521–542, 1996.

[CRA 12] CRAVEN A., Electromobility in Sweden: facilitating market conditions to encourage consumer uptake of electric vehicles, Masters Thesis, Uppsala University, 2012.

[DAV 85] DAVID P.A., "Clio and the economics of QWERTY", *The American Economic Review*, vol. 75, no. 2, pp. 332–337, 1985.

[DAV 14] DAVID B., DE LATTRE-GASQUET M., MATHY S. *et al.*, "Prospective énergétique: le possible, le souhaitable et l'acceptable", *Futuribles*, no. 398, p. 37, January–February 2014.

[DEB 13] DEBEIR J.-C., DELEAGE J.-P., HEMERY D., *Une histoire de l'énergie. Les servitudes de la puissance*, Flammarion, Paris, 2013.

[DEL 12a] DELANG C.O., CHENG W.T., "Consumers' attitudes towards electric cars: a case study of Hong Kong", *Transportation Research Part D*, vol. 17, pp. 492–494, 2012.

[DEL 12b] DELFOSSE C., GUIOMAR X., PLUVINAGE J. *et al.* (eds), "Dossier alimentation et territoires", *Revue POUR*, nos. 215–216, pp. 57–383, 2012.

[DEP 09] DEPRET M., HAMDOUCH A., "Quelles politiques de l'innovation et de l'environnement pour quelle dynamique d'innovation environnementale?", *Innovations. Cahiers d'Économie de l'Innovation*, no. 29, pp. 127–147, 2009.

[DES 07] DESCOTES A., MONCOMBLE D., Bilan carbone et plan climat de la Champagne: du diagnostic au passage à l'acte, Comité Interprofessionnel du Vin de Champagne (CIVC) Épernay, France, 2007.

[DEV 10] DEVERRE C., LAMINE C., "Les systèmes agroalimentaires alternatifs – Une revue de travaux anglophones en sciences sociales", *Economie Rurale*, no. 317, pp. 57–73, 2010.

[DEW 11] DEWALD U., TRUFFER B., "Market formation in technological innovation systems – diffusion of photovoltaic applications in Germany", *Industry and Innovation*, vol. 18, no. 3, pp. 285–300, 2011.

[DEW 12] DEWALD U., TRUFFER B., "The local sources of market formation: explaining regional growth differentials in German photovoltaic markets", *European Planning Studies*, vol. 20, no. 3, pp. 397–420, 2012.

[DIE 07] DIEMER A., LABRUNE S., "L'écologie industrielle: quand l'écosystème industriel devient un vecteur du développement durable", *Développement durable et territoires*, Varia, 30 August 2007.

[DOS 82] DOSI G., "Technological paradigms and technological trajectories: a suggested interpretation of the determinants and directions of technical change", *Research Policy*, vol. 11, no. 3, pp. 147–162, 1982.

[DU 11] DU TERTRE C., "Modèles économiques d'entreprise, dynamique macroéconomique et développement durable", in GAGLIO G., LAURIOL J., DU TERTRE C. (eds), *L'économie de la fonctionnalité: une voie nouvelle vers un développement durable?*, Octarès, Toulouse, pp. 21–42, 2011.

[DUD 14] DUDENHOFFER K., ARORA R., DIVERREZ A. *et al.*, Potentials for electric vehicles in France, Germany, and India, Working paper series in production and energy, Karlsruhe Institute of Technology, 2014.

[DUR 14a] DURU M., FARES M., THEROND O., "Un cadre conceptuel pour penser maintenant (et organiser demain) la transition agroécologique de l'agriculture dans les territoires", *Cahiers Agricultures*, vol. 23, no. 2, pp. 84–95, 2014.

[DUR 14b] DURUISSEAU K., "L'émergence du concept de transition énergétique. Quels apports de la géographie?", *Bulletin de la Société Géographique de Liège*, no. 63, pp. 21–34, 2014.

[DUT 13] DUTSCHKE E., SCHNEIDER U., PETERS A., Who will use electric vehicles?, Working paper sustainability and innovation, no. S6, 2013.

[EDE 06] EDENHOFER O., CARRARO C., KÖHLER J., "Endogeneous technological change and the economics of atmospheric stabilisation", *Energy Journal*, vol. 27, Special issue, 2006.

[EHR 00] EHRENFELD J.R., "Industrial ecology: paradigm shift or normal science?", *American Behavioral Scientist*, vol. 44, no. 2, pp. 229–244, 2000.

[ELZ 04] ELZEN B., GEELS F.W., GREEN K. (eds), *System Innovation and the Transition to Sustainability. Theory, Evidence and Policy*, Edward Elgar, Cheltenham, 2004.

[EME 07] EMELIANOFF C., "La ville durable: l'hypothèse d'un tournant urbanistique en Europe", *L'information géographique*, vol. 71, no. 3, pp. 48–65, 2007.

[ENS 13] ENSSLEN A., JOCHEM P., FICHTNER W., "Experiences of EV users in the French–German context", *Proceedings of EVS*, no. 27, pp. 1–12, 2013.

[ERK 98] ERKMAN S., *Vers une écologie industrielle: comment mettre en pratique le développement durable dans une société hyper-industrielle*, Editions Charles Léopold Mayer, 1998.

[EU 10] EU COMMISSION TASK FORCE FOR SMART GRIDS, Expert Group 1: Functionalities of smart grids and smart meters, available at: https://ec.europa.eu/energy/en/topics/markets-and-consumers/smart-grids-and-meters/smart-grids-task-force, 2010.

[EUR 05] EUROSTAT, *Manuel d'Oslo: principes directeurs pour le recueil et l'interprétation des données sur l'innovation*, OECD Publishing, 2005.

[FAR 10] FARLA J, ALKEMADE F., SUURS R.A.A., "Analysis of barriers in the transition towards sustainable mobility in the Netherlands", *Technological Forecasting and Social Change*, vol. 77, no. 8, pp. 1260–1269, 2010.

[FAR 12] FARES M., MAGRINI M.B., TRIBOULET P., "Transition agroécologique, innovation et effets de verrouillage: le rôle de la structure organisationnelle des filières – Le cas de la filière blé dur française", *Cahiers de l'Agriculture*, vol. 21, no. 1, pp. 34–45, 2012.

[FIN 10] FINON D., DAMIAN M., "Le captage et le stockage du carbone: entre nécessité et réalisme", *Cahier de Recherche LEPII*, no. 40, December 2010.

[FIS 90] FISCHLER C., *Homnivore (L'): Sur les fondamentaux de la biologie et de la philosophie*, Editions Odile Jacob, 1990.

[FLI 09] FLIPO F., GOSSART C., "Infrastructure numérique et environnement: l'impossible domestication de l'effet rebond", *Terminal*, nos. 103–104, pp. 163–178, 2009.

[FOX 08] FOXON T.J., KÖHLER J., OUGHTON C. (eds), *Innovation for a Low Carbon Economy. Economic Institutional and Management Approaches*, Edward Elgar, 2008.

[FRA 10] FRANTZESKAKI N., LOORBACH D., "Towards governing infrasystem transitions: reinforcing lock-in or facilitating change?", *Technological Forecasting and Social Change*, vol. 77, no. 8, pp. 1292–1301, 2010.

[FRE 09] FREYSSENET M., "La voiture électrique: le début d'une deuxième révolution automobile?", *Journée du Gerpisa*, no. 176, December 2009.

[FRO 89] FROSCH R.A., GALLOPOULOS N.E., "Strategies for manufacturing", *Scientific American*, vol. 261, no. 3, pp. 144–152, 1989.

[FUS 96] FUSSLER C., JAMES P., *Driving Eco-Innovation, A Breakthrough Discipline for Innovation and Sustainability*, Pitman Publishing, 1996.

[GAL 12] GALLAUD D., MARTIN M., REBOUD S. *et al.*, "La relation entre innovation environnementale et réglementation – Une application au secteur agroalimentaire français", *Innovations. Cahiers d'Économie de l'Innovation*, vol. 1, no. 3, pp. 155–175, February–March 2012.

[GAL 13] GALLIANO D., NADEL S., "Les déterminants de l'adoption de l'éco-innovation selon le profil stratégique de la firme: le cas des firmes industrielles françaises", *Revue d'économie industrielle*, vol. 142, pp. 77–110, 2013.

[GAL 15] GALLIANO D., RAYNAUD E., Document de synthèse sur les travaux du département portant sur les filières et sur les systèmes alimentaires durables, SAD, October, p. 46, 2015.

[GAL 16] GALLAUD D., LAPERCHE B., *Circular Economy, Industrial Ecology and Short Supply Chain*, ISTE Ltd, London and John Wiley & Sons, New York, 2016.

[GAS 03] GASMI N., GROLLEAU G., "Spécificités des innovations environnementales: Une application aux systèmes agro-alimentaires", *Innovations. Cahiers d'Économie de l'Innovation*, vol. 18, pp. 73–89, 2003.

[GEE 02] GEELS F.W., "Technological transitions as evolutionary reconfiguration processes: a multi-level perspective and a case study", *Research Policy*, vol. 31, nos. 8–9, pp. 1257–1274, 2002.

[GEE 04] GEELS F.W., ELZEN B., GREEN K., "General introduction: system innovation and transitions to sustainability", in ELZEN B., GEELS F., GREEN K. (eds), *Systems Innovation and the Transition to Sustainability: Theory, Evidence and Policy*, Edward Elgar, Cheltenham, pp. 1–16, 2004.

[GEE 05a] GEELS F.W., "Co-evolution of technology and society: the transition in water supply and personal hygiene in the Netherlands (1850–1930): a case study in multi-level perspective", *Technology in Society*, vol. 27, no. 3, pp. 363–397, 2005.

[GEE 05b] GEELS F.W., *Technological Transitions and System Innovations. A Co-evolutionary and Socio-technical Analysis*, Edward Elgar, Cheltenham, 2005.

[GEE 06a] GEELS F.W., "Multi-level perspective on system innovation: relevance for industrial transformation", in OLSHOERN X., WIECZOREK J. (eds), *Understanding Industrial Transformation: Views from Different Disciplines*, Springer, pp. 163–186, 2006.

[GEE 06b] GEELS F.W., "The hygienic transition from cesspools to sewer systems (1840–1930): the dynamics of regime transformation", *Research Policy*, vol. 35, pp. 1069–1082, 2006.

[GEE 07] GEELS F.W., SCHOT J., "Taxonomy of transition pathways in socio-technical systems", *Research Policy*, vol. 36, pp. 399–417, 2007.

[GEE 10] GEELS F.W., SCHOT J., "The dynamics of sociotechnical transitions: a socio-technical perspective", in GRIN J., ROTMANS J., SCHOT J. (eds), *Transitions to Sustainable Development: New Directions in the Study of Long Term Transformative Change*, Routledge, New York, pp. 9-101, 2010

[GEE 12] GEELS F.W., KEMP R., DUDLEY G. *et al.* (eds), *Automobility in Transition? A Socio-technical Analysis of Sustainable Transport*, Routledge, New York, 2012.

[GEN 08] GENUS A., COLES A.-M., "Rethinking the multi-level perspective of technological transitions", *Research Policy*, vol. 37, pp. 1436–1445, 2008.

[GER 12a] GERPISA, La voiture électrique comme artefact d'une transition vers une économie écologique, Report, available at: http://gerpisa.org/system/files/Rapport_final_VE_projet_TEE_0.pdf, 2012.

[GER 12b] GERPISA, Le plan véhicule vert Bretagne, une tentative territoriale pour favoriser une transition vers une économie écologique, Report, available at: http://gerpisa.org/system/files/TEE_PartieVVB_Corps_du_rapport-1_0.pdf, 2012.

[GID 84] GIDDENS A., *The Constitution of Society*, Polity Press, Cambridge, 1984.

[GOU 10] GOULET F., "Nature et ré-enchantement du monde", in HERVIEU B., MAYER N., MULLER P. *et al.* (eds), *Les mondes agricoles en politique*, Presses de Sciences Po, Paris, pp. 51–72, 2010.

[GRI 10] GRIN J., ROTMANS J., SCHOT J., *Transitions to Sustainable Development. New Directions in the Study of Long Term Transformative Change*, Routledge, 2010.

[GRU 02] GRUBB M., KÖHLER J., ANDERSON D., "Induced technical change in energy and environmental modeling: analytic approaches and policy implications", *Annual Review of Energy and Environment*, vol. 27, pp. 271–308, 2002.

[GRU 04] GRUBB M., "Technology innovation and climate change policy: an overview of issues and options", *Keio Economic Studies*, vol. 41, no. 2, pp. 103–132, 2004.

[GUY 13] GUYOMARD H. (ed.), *Vers des agricultures à hautes performances*, "Comment rendre l'Agriculture Biologique française plus productive et plus competitive", vol. 1, INRA, p. 36, 2013.

[HA 10] HA-DUONG M., CHAABANE N. (eds), *Le captage et le stockage du CO_2. Enjeux techniques et sociaux en France*, Quae, 2010.

[HAL 10] HALLEGATTE S., DE PERTHUIS C., LECOCQ F., Economie de l'adaptation au changement climatique, Conseil Economique pour le Développement Durable, February, 2010.

[HAN 10] HANSEN J.P., PERCEBOIS J., BOITEUX M. *et al.*, *Energie: Economie et politique*, De Boeck, Brussels, 2010.

[HAN 14] HANKE C., HULSMANN M., FORNAHL D., "Socio-economic aspects of electric vehicles: a literature review", in HÜLSMANN M., FORNAHL D. (eds), *Evolutionary Paths Towards the Mobility Patterns of the Future*, Springer, Berlin, Heidelberg, pp. 13–36, 2014.

[HEK 07] HEKKERT M.P., SUURS R.A.A., NEGRO S.O. *et al.*, "Functions of innovation systems: a new approach for analyzing technological change", *Technological Forecasting and Social Change*, vol. 74, pp. 413–432, 2007.

[HEK 11] HEKKERT M.P., NEGRO S.O., HARMSEN R. *et al.*, Technological innovation system analysis: a manual for analysts, Report, Energy Institute, Utrecht University, The Netherlands, 2011.

[HEL 10] HELLSMARK H., Unfolding the Formative Phase of Gasified Biomass in the European Union: the Role of System Builders in Realizing the Potential of Second-generation Transportation Fuels from Biomass, Chalmers University of Technology, Gothenburg, Sweden, 2010.

[HEN 16] HENRIET D., EXTERNALITÉ, économie, Encyclopedia Universalis, available at: http://www.universalis.fr/encyclopedia/externalite.economie, 2016.

[HET 07] HETLAND J., MULDER G., "In search of a sustainable hydrogen economy: how a large-scale transition to hydrogen may affect the primary energy demand and greenhouse gas emissions", *International Journal of Hydrogen*, vol. 32, pp. 736–747, 2007.

[HEY 14] HEYMANS C., WALKER S.B., YOUNG S.B. *et al.*, "Economic analysis of second use electric vehicle batteries for residential energy storage and load-levelling", *Energy Policy*, vol. 71, pp. 22–30, 2014.

[HIS 06] HISSCHEMOLLER M., BODE R., VAN DE KERHOF M., "What governs the transition to a sustainable hydrogen economy? Articulating the relationship between technologies and political institutions", *Energy Policy*, vol. 34, pp. 1227–1235, 2006.

[HOO 02] HOOGMA, R., KEMP R., SCHOT J. *et al.*, *Experimenting for Sustainable Transport: the Approach of Strategic Niche Management*, Spon Press, London and New York, 2002.

[HOP 10] HOPKINS R., *Manuel de transition: de la dépendance au pétrole à la résilience locale*, Editions Ecosociété, 2010.

[HOT 31] HOTELLING H., "The economics of exhaustible resources", *The Journal of Political Economy*, vol. 39, no. 2, pp. 131–175, 1931.

[HUG 69] HUGHES T.P., "Technological momentum in history: hydrogenation in Germany 1898–1933", *Past and Present*, vol. 44, pp. 106–132, 1969.

[INT 00] INTERNATIONAL ENERGY AGENCY (IEA), *Experience Curves for Technology Policies*, OECD Publishing, 2000.

[INT 08a] INTERNATIONAL ENERGY AGENCY (IEA), *CO_2 Capture and Storage: a Key Carbon Abatement Option*, OECD Publishing, 2008.

[INT 08b] INTERNATIONAL ENERGY AGENCY (IEA), *Energy Technology Perspectives*, OECD Publishing, p. 646, 2008.

[INT 11] INTERNATIONAL ENERGY AGENCY (IEA), *Technology Roadmap Smart Grids*, OECD Publishing, 2011.

[INT 13] INTERNATIONAL ENERGY AGENCY (IEA), *World Energy Outlook Special Report 2013: Redrawing the Energy Climate Map*, OECD, Publishing, p. 132, 2013.

[INT 15] INTERNATIONAL ENERGY AGENCY (IEA), *Energy Technology Perspectives*, OECD Publishing, 2015.

[JAB 12] JABEEN F., OLARU D., SMITH B. *et al.*, "Acceptability of electric vehicles: Findings from a driver survey", *Transport Research Board of the National Academics* 2012 ATRF Papers, pp. 1–15, 2012.

[JAC 00] JACOBSSON S., JOHNSON A., "The diffusion of renewable energy technology: an analytical framework and key issues for research", *Energy Policy*, vol. 28, pp. 625–640, 2000.

[JAC 03] JACOBSSON S., BERGEK A., Energy systems transformation: The evolution of technological systems in renewable energy technology, Department of Industrial Dynamics, Chalmers University of Technology, Gothenburg, Sweden, 2003.

[JAC 04a] JACOBSSON S., BERGEK A., "Transforming the energy sector: the evolution of technological systems in renewable energy technology", *Industrial and Corporate Change*, vol. 13, no. 5, pp. 815–849, 2004.

[JAC 04b] JACOBSSON S., SANDEN B., BANGENS L., "Transforming the energy system – the evolution of the German technological system for solar cells", *Technological Analysis and Strategic Management*, vol. 16, no. 1, pp. 3–30, 2004.

[JAC 06] JACOBSSON S., LAUBER V., "The politics and policy of energy system transformation – explaining the diffusion of renewable energy technology", *Energy Policy*, vol. 34, pp. 256–276, 2006.

[JAC 08] JACOBSSON S., "The emergence and troubled growth of a 'biopower' innovation system in Sweden", *Energy Policy*, vol. 36, no. 4, pp. 1491–1508, 2008.

[JAC 11] JACOBSSON S., BERGEK A., "Innovation system analyses and sustainability transitions: Contributions and suggestions for research" *Environmental Innovation and Societal Transitions*, vol. 1, no. 1, pp. 41–57, 2011.

[JAN 13] JANCOVICI J.-M., "Quelques millénaires de transition énergétique", *La Jaune et la Rouge*, no. 689, 2013.

[JOH 98] JOHNSON A., "Functions in innovation system approaches", Working paper, Department of Industrial Dynamics, Chalmers University of Technology, Gothenburg, 1998.

[JOH 01] JOHNSON A., "Functions in innovation system approaches", in *Nelson and Winter Conference*, 12–15 June, Aalborg, Denmark, p. 19, 2001.

[JUL 12] JULLIEN B., VILLAREAL A., "La voiture électrique comme artefact d'une transition vers une économie écologique?", Report, GERPISA, December, p. 34, 2012.

[KAL 13] KALINOWSLI L., PASTOR J.M., "L'hydrogène: vecteur de la transition énergétique?", Report, OPECST No. 1672, December, p. 50, 2013.

[KAP 09] KAPLAN F., *La métamorphose des objets*, FYP Editions, 2009.

[KAU 08] KAUFMANN V., "*Les paradoxes de la mobilité. Bouger, s'enraciner*", Presses polytechniques et universitaires romandes, 2008.

[KEM 94] KEMP R., "Technology and the transition to environmental sustainability: the problem of technological regime shifts", *Futures*, vol. 26, no. 10, pp. 1023–1046, 1994.

[KEM 98] KEMP R., SCHOT J., HOOGMA R., "Regime shifts to sustainability through processes of niche formation: the approach of strategic niche management", *Technology Analysis and Strategic Management*, vol. 2, no. 10, pp. 175–195, 1998.

[KEM 01] KEMP R., RIP A., SCHOT J.W., "Constructing transition paths through the management of niches", in GARUD R., KARNOE P. (eds), *Path Dependence and Creation*, Lawrence Erlbaum, Mahwah, pp. 269–299, 2001.

[KEM 05] KEMP R., ROTMANS J., "The management of the co-evolution of technical environmental and social systems", in WEBER M., HEMMELSKAMP J. (eds), *Towards Environmental Innovation Systems*, Springer, 2005.

[KEM 10] KEMP R., "Eco-innovation: definition, measurement and open research issues", *Economia Politica*, vol. 27, pp. 397–420, 2010.

[KLE 11] KLEY F., LERCH C.H., DALLINGER D., "New business models for electric cars. A holistic approach", *Energy Policy*, vol. 39, pp. 3392–3403, 2011.

[KLE 12] KLERKX L., VAN MIERLO B., LEEUWIS C., "Evolution of systems approaches to agricultural innovation: concepts, analysis and interventions", in DARNHOFER I., GIBBON D., DEDIEU B. (eds), *Farming Systems Research into the 21st Century: The New Dynamic,* Springer, Dordrecht, pp. 457–483, 2012.

[KLI 12] KLITKOU A., BORUP M., Energy Innovation Systems Indicator Report, DTU Management Engineering and NIFU, Nordic Institute for Studies in Innovation, Research and Education, 2012.

[KOH 09] KOHLER J., WHITMARSH L., NYKVIST B. *et al.*, "A transition model for sustainable mobility", *Ecological Economics*, vol. 68, pp. 2985–2995, 2009.

[KON 12] KONRAD K., MARKARD J., RUEF A. *et al.*, "Strategic responses to fuel cell hype and disappointment", *Technological Forecasting and Social Change*, vol. 79, no. 6, pp. 1084–1098, 2012.

[LAF 06] LAFFITTE P., SAUNIER C., "Les apports de la science et de la technologie au développement durable, Tome I: Changement climatique et transition énergétique: dépasser la crise", Report, OPECST, 2006.

[LAM 08] LAMINE C., PERROT N., *Les AMAP: un nouveau pacte entre producteurs et consommateurs?,* Editions Yves Michel, 2008.

[LAM 09a] LAMINE C., VIAUX P., MORIN J.M. *et al.*, "Dynamiques de développement de l'agriculture biologique: éléments de débat", *Innovations agronomiques*, vol. 4, pp. 483–493, 2009.

[LAM 09b] LAMINE C., MEYNARD J.M., PERROT N. *et al.*, Analyse des formes de transition vers des agricultures plus écologiques: les cas de l'Agriculture Biologique et de la Protection Intégrée, *Innovations Agronomiques*, vol. 4, pp. 483–493, 2009.

[LAM 10] LAMINE C., BELLON S., *Transitions vers l'agriculture biologique*, Educagri Editions, Dijon, 2010.

[LAM 12a] LAMINE C., "Changer de système: une analyse des transitions vers l'agriculture biologique à l'échelle des systèmes agri-alimentaires territoriaux", *Terrains & travaux*, vol. 1, pp. 139–156, 2012.

[LAM 12b] LAMINE C., CHIFFOLEAU Y., "Reconnecter agriculture et alimentation dans les territoires: dynamiques et défis", *Pour*, vol. 3, nos. 215–216, pp. 85–92, 2012.

[LAS 15] LASBLEIZ R., STOKKINK D., L'agroécologie: Inscrire l'agriculture dans la transition, Note d'analyse, Pour la solidarité, 2015.

[LAT 13] LATRUFFE L., NAUGES C., DESJEUX Y., Motivations et freins à la conversion en agriculture biologique des producteurs laitiers et légumiers: Résultats d'une enquête à grande échelle en Bretagne et en Pays de la Loire, Deliverable 2, AgriBio3 PEPP Research Project, p. 91, 2013.

[LE 15] LE DEAUT J.-Y., SIDO B., Innovation et changement climatique: l'apport de l'évaluation scientifique et technologique, Report, OPECST, no. 3206, November 2015.

[LEC 10] LECOURTIER J., "Préambule – Le stockage du CO_2 au service de la lutte contre le changement climatique", *Oil & Gas Science and Technology – Rev. IFP*, vol. 65, no. 4, pp. 529–532, 2010.

[LEP 06] LEPOUTRE J., HEENE A., "Investigating the impact of firm size on small business social responsibility: a critical review", *Journal of Business Ethics*, vol. 67, no. 3, pp. 257–273, 2006.

[LES 11] LESIEUR C., "L'agroalimentaire innove en faveur de l'environnement", *Agreste Primeur*, no. 269, October, p. 4, 2011.

[LIE 95] LIEBOWITZ S.J., MARGOLIS S.E., "Path dependence, lock-in, and history". *Journal of Law, Economics and Organization*, vol. 11, pp. 205–226, 1995.

[LIE 11] LIEVEN T., MUHLMEIER S., HENKE S. *et al.*, "Who will buy electric cars? An empirical study in Germany", *Transportation Research Part D*, vol. 16, pp. 236–243, 2011.

[LOO 07] LOORBACH D., *Transition Management: New Mode of Governance for Sustainable Development*, International Books, Utrecht, p. 327, 2007.

[MAG 13] MAGRINI M.B., TRIBOULET P., BEDOUSSAC L., "Pratiques agricoles innovantes et logistique des coopératives agricoles – Une étude ex-ante sur l'acceptabilité de cultures associées blé dur – légumineuses", *Economie Rurale*, vol. 38, pp. 25–45, 2013.

[MAL 47] MALTHUS T.R., *Principes d'économie politique considérés sous le rapport de leur application pratique*, Guillaumin, 1847.

[MAR 06] MARKARD J., TRUFFER B., "Innovation processes in large technical systems: Market liberalization as a driver for radical change?", *Research Policy*, vol. 35, no. 5, pp. 609–625, 2006.

[MAR 08a] MARKARD J., TRUFFER B., "Actor-oriented analysis of innovation systems: exploring micro-meso level linkages in the case of stationary fuel cells", *Technological Analysis Strategic Management*, vol. 20, pp. 443–464, 2008.

[MAR 08b] MARKARD J., TRUFFER B., "Technological innovation systems and the multi-level perspective: towards an integrated framework", *Research Policy*, vol. 37, pp. 596–615, 2008.

[MAR 09] MARKARD J., PETERSEN R., "The offshore trend: structural changes in the wind power sector", *Energy Policy*, vol. 37, pp. 3545–3556, 2009.

[MAR 10] MARECHAL G., SPANU A., "Les circuits courts favorisent-ils l'adoption de pratiques agricoles plus respectueuses de l'environnement?", *Le Courrier de l'environnement de l'INRA*, vol. 59, no. 59, pp. 33–45, 2010.

[MAR 12] MARKARD J., RAVEN R., TRUFFER B., "Sustainability transitions: an emerging field of research and its prospects", *Research Policy*, vol. 41, pp. 955–967, 2012.

[MAS 09] MASERO J., "Investissements et études pour protéger l'environnement dans l'agroalimentaire, reprise des investissements antipollution", *Agreste Primeur*, November, no. 232, p. 4, 2009.

[MAS 10a] MASERO J., "Protection de l'environnement dans l'industrie agroalimentaire. Les investissements pour le traitement des eaux usées résistent bien en 2009", *Agreste Primeur*, November, no. 249, p. 4, 2010.

[MAS 10b] MASERO J., "Déchets des industries agroalimentaires. Une bonne gestion des déchets organiques", *Agreste Primeur*, July, no. 245, p. 4, 2010.

[MEA 72] MEADOWS D.H., MEADOWS D.L., RANDERS J. *et al.*, "Rapport sur les limites de la croissance", in DELAUNAY J. (ed.) *Halte à la croissance?*, Fayard, Paris, 1972.

[MEI 15] MEISSONNIER J., "Le déménagement comme occasion de rompre l'inertie des routines de mobilité quotidienne et de s'engager dans une mobilité durable? Décryptage d'un paradoxe", in SCARWELL H. J., LEDUCQ D., GROUX A. *et al.*, *Réussir la transition énergétique*, Presses universitaires du Septentrion. pp. 133–140, 2015.

[MEL 08] MELAINA M., BREMSON J., "Refueling availability for alternative fuel vehicle markets: sufficient urban station coverage", *Energy Policy*, vol. 36, no. 8, pp. 3233–3241, 2008.

[MEY 13a] MEYNARD J.M., MESSÉAN A., CHARLIER A. *et al.*, *Freins et leviers à la diversification des cultures. Étude au niveau des exploitations agricoles et des filières*, Éditions Inra, p. 52, 2013.

[MEY 13b] MEYNARD J.M., MESSÉAN A., CHARLIER A. *et al.*, "Freins et leviers à la diversification des cultures: étude au niveau des exploitations agricoles et des filières", *Oléagineux Corps Gras Lipides*, vol. 20, no. 4, p. D403, 2013.

[MIL 14] MILLOCK K., NAUGES C., "Comportement des ménages et consommation alimentaire", in OCDE (ed.), *Vers des comportements plus environnementaux: Vue d'ensemble de l'enquête 2011*, OECD Publishing, Paris, pp. 203–240, 2014.

[MIS 14] MISSEMER A., L'analyse économique face à l'épuisement des ressources naturelles, de William Stanley Jevons à Harold Hotelling (1865–1931) – le cas des énergies fossiles, Thesis, University Lyon 2, 2014.

[MOL 00] MOL A., SONNENFELD D., *Ecological Modernization Around the World. Perspectives and Critical Debates*, Frank Cass, London, 2000.

[MON 02] MONT O., "Clarifying the concept of Product-Service System", *Journal of Cleaner Production*, vol. 10, no. 3, pp. 237–245, 2002.

[MOR 16] MOREL A.L., LE ROUX S., *Fab Labs: Innovative User*, ISTE Ltd, London and John Wiley & Sons, New York, 2016.

[MOU 15] MOUGENOT B., "Automobile et nouveaux modèles économiques de la mobilité électrique, au cœur d'une diversité institutionnelle", *Innovations*, vol. 1, no. 46, pp. 71–88, 2015.

[MUS 12] MUSIOLIK J., MARKARD J., "Networks and network resources in technological innovation system. Towards a conceptual framework for a system building", *Technological Forecasting and Social Change*, vol. 79, no. 6, pp. 1032–1048, 2012.

[NAT 15] NATIONAL SCIENCE FOUNDATION, "Improving electric vehicles sales may require solving unique chicken and egg problem, article, National Science Foundation, available at: www.nsf.gov/discoveries/dics_summ.jsp?cntn_id=133947, 2015.

[NEG 07] NEGRO S.O., HEKKERT M.P., SMITS R.E., "Explaining the failure of the Dutch innovation system for biomass digestion – a functional analysis", *Energy Policy*, vol. 35, no. 2, pp. 925–938, 2007.

[NEG 08] NEGRO S.O., HEKKERT M.P., "Explaining the success of emerging technologies by innovation system functioning: the case of biomass digestion in Germany", *Technology Analysis & Strategic Management*, vol. 20, no. 4, pp. 465–482, 2008.

[NEL 82] NELSON R.R., WINTER S.G., *An Evolutionary Theory of Economic Change*, Belknap Press/Harvard University Press, Cambridge, MA, 1982.

[NIE 06] NIEUWENHUIS P., VERGRAGT P., WELLS P. (eds), *The Business of Sustainable Mobility: from Vision to Reality*, Greenleaf, Sheffield, 2006.

[NIL 09] NILL J., KEMP R., "Evolutionary approaches for sustainable innovation policies: from niche to paradigm?", *Research Policy*, vol. 38, pp. 668–680, 2009.

[OCD 09] OCDE, *Politique de l'environnement et innovation technologique*, Editions OCDE, Paris, 2009.

[OCD 10] OCDE, *L'éco-innovation, dans l'industrie. Favoriser la croissance verte*, Editions OCDE, Paris, 2010.

[OCD 11] OCDE, *Vers une croissance verte*, Editions OCDE, Paris, 2011.

[OEC 10] OECD, Eco-Innovation in Industry: Enabling Green Growth, OECD Publishing, 2010.

[OEC 11] OECD, *Towards Green Growth*, OECD Publishing, 2011.

[ONU 15] ONU, World Population Prospects: The 2015 Revision, Key Findings & Advance Tables, United Nations, working paper no. ESA/P/WP.241, 2015.

[PER 15] PERCEBOIS J., MANDIL C. (eds), *Rapport Energie 2050, Centre d'Analyse Stratégique*, La Documentation française, Centre d'analyse stratégique, 2015.

[PIC 12] PICARD F., "Open innovation and joint patent applications: the case of greenhouse gas capture and storage technologies", *Journal of Innovation Economics & Management*, vol. 2, no. 10, pp. 107–122, 2012.

[PIS 13] PISON G. (ed.), "The population of the world", *Population and Societies*, no. 503, September, 2013.

[PON 13] PONSOT-JACQUIN C., BERTRAND J.F., Panorama 2013: Le stockage massif de l'énergie, IFN-EN, 2013.

[POR 95] PORTER M.E., VAN DER LINDE C., "Toward a new conception of the environmental competitiveness relationship", *Journal of Economic Perspectives*, vol. 9, no. 4, pp. 97–118, 1995.

[POR 06] PORTER M.E., KRAMER M.R., "Strategy and society: the link between competitive advantage and corporate social responsibility", *Harvard Business Review*, vol. 84, pp. 78–92, 2006.

[POS 96] POSSAS M.L., SALLES-FILHO S., DA SILVEIRA J., "An evolutionary approach to technological innovation in agriculture: some preliminary remarks", *Research Policy*, vol. 25, no. 6, pp. 933–945, 1996.

[POU 11] POUPAT B., "Des pratiques environnementales, sous contraintes, Les perceptions socials et pratiques environnementales des Français de 1995 à 2011", *Revue du CGDD*, October 2011.

[QUA 07] QUAIREL F., AUBERGER M.N., "La diffusion de la RSE par la relation fournisseurs. Injonctions paradoxales ou partenariat de progrès?", *Revue Internationale des PME*, vol. 20, nos. 3–4, pp. 69–94, 2007.

[QUI 13] QUITZOW R., "Towards an integrated approach to promoting environmental innovation and national competitiveness", *Innovation and Development*, vol. 3, no. 2, pp. 277–296, 2013.

[RAR 11] RARE., (RÉSEAU DES AGENCES RÉGIONALES DE L'ENERGIE ET DE L'ENVIRONNEMENT), "Typologie des facteurs d'émission", *Fiche Ressource*, no. 3, June 2011.

[RAV 12] RAVEN R., SCHOT J., BERKOUT F., "Space and scale in socio-technical transitions", *Environmental Innovation and Societal Transitions*, vol. 4, pp. 63–78, 2012.

[RED 06] REDLINGSHÖFER B., "Vers une alimentation durable? Ce qu'enseigne la littérature scientifique", *Le courrier de l'environnement de l'INRA*, vol. 53, pp. 83–102, 2006.

[REN 00] RENNINGS K., "Redefining innovation – eco-innovation research and the contribution from ecological economics", *Ecological Economics*, vol. 32, pp. 319–332, 2000.

[RIC 04] RICARDO D., *On the Principles of Political Economy and Taxation*, Dover Publications, 2004.

[RIF 02] RIFKIN J., *L'économie hydrogène. Après la fin du pétrole, la nouvelle révolution économique*, La Découverte, 2002.

[RIF 12] RIFKIN J., *La troisième révolution industrielle. Comment le pouvoir latéral va transformer l'énergie, l'économie et le monde*, Les liens qui libèrent, Paris, 2012.

[RIP 95] RIP A., "Introduction of new technology: making use of recent insights from sociology and economics of technology", *Technology Analysis and Strategic Management*, vol. 7, no. 4, pp. 417–432, 1995.

[RIP 98] RIP A., KEMP R., "Technological change", in RAYNER S., MALONE E.L. (eds), *Human Choice and Climate Change*, Battelle Press, Columbus, pp. 327–399, 1998.

[RIZ 10] RIZET C., BROWNE M., CORNELIS E. *et al.*, "Emissions de gaz à effet de serre des chaînes logistiques", *Logistique et Management*, vol. 18, no. 2, pp. 73–83, 2010.

[ROM 04] ROMM J.J., *The hype about hydrogen. Fact and fiction in the race to save the climate*, Island Press, 2004.

[ROT 01] ROTMANS J., KEMP R., VAN ASSELT M., "More evolution than revolution: transition management in public policy", *Foresight*, vol. 3, no. 1, pp. 15–31, 2001.

[ROT 10] ROTILLON G., *Economie des ressources naturelles*, La Découverte, 2010.

[ROU 14] ROUQUET R., Comparaison internationale des politiques publiques en matière d'économie circulaire, General Commission for Sustainable Development, no. 101, p. 57, January 2014.

[RUM 13] RUMPALA Y., "Formes alternatives de production énergétique et reconfigurations politiques. La sociologie des énergies alternatives comme études des potentialités de réorganisation du collectif", *Flux*, no. 92, pp. 47–61, 2013.

[SAC 77] SACHS I., *Pour une économie politique du développement*, Flammarion, Paris, 1977.

[SCA 15] SCARWELL H.-J., LEDUCQ D., GROUX A. (eds), *Réussir la transition énergétique*, Presses Universitaires du Septentrion, Lille, 2015.

[SCH 35] SCHUMPETER J., *The Theory of Economic Development*, Transaction Publishers, 1935.

[SCH 97] SCHOT J., RIP A., "The past and future of constructive technology assessment", *Technological Forecasting and Social Change*, vol. 54, no. 2, pp. 251–268, 1997.

[SCH 12] SCHLÖR H., FISCHER W., HAKE J.F., "The meaning of energy systems for the genesis of the concept of sustainable development", *Applied Energy*, vol. 97, pp. 192–200, 2012.

[SCH 15] SCHNEIDER A., HUYGHE C. (eds), *Les légumineuses pour des systèmes agricoles et alimentaires durables*, Editions Quae, Versailles, p. 473, 2015.

[SDS 15] SDSN – IDDRI, "Pathways to deep decarbonization 2015 report – executive summary", Deep Decarbonization Pathways Project, SDSN – IDDRI, 2015.

[SID 13] SIDO B., LE DEAUT J.-Y., La transition énergétique à l'aune de l'innovation et de la décentralisation, Report, OPECST, no. 1352, September 2013.

[SMI 05] SMITH A., STIRLING A., BERKHOUT F., "The governance of sustainable socio-technical transitions", *Research Policy*, vol. 34, no. 10, pp. 1491–1510, 2005.

[SOC 96] SOCOLOW R., ANDREWS C., BERKHOUT F. *et al.*, *Industrial Ecology and Global Change*, Cambridge University Press, 1996.

[SOL 10] SOLIER B., TROTIGNON R., *Comprendre les enjeux énergétiques*, Pearson Education, 2010.

[SOV 10] SOVACOOL B.K., BROSSMANN B., "Symbolic convergence and the hydrogen economy", *Energy Policy*, vol. 38, pp. 1999–2012, 2010.

[STA 12] STASSART P.M., BARET P.H., GRÉGOIRE J.-C.L. *et al.*, "L'agroécologie: trajectoire et potentiel – Pour une transition vers des systèmes alimentaires durables", in VAN DAM D., NIZET J., STREITH M. *et al.* (eds), *Agroécologie entre pratiques et sciences sociales*, Educagri éditions, pp. 25–51, 2012.

[STE 07] STERN N., *The Economics of Climate Change: The Stern Review*, Cambridge University Press, 2007.

[SUU 09a] SUURS R.A.A., Motors of sustainable innovation. Towards a theory on the dynamics of technological innovation systems, Thesis, Utrecht University, Utrecht, 2009.

[SUU 09b] SUURS R.A., HEKKERT M.P., "Cumulative causation in the formation of a technological innovation system: the case of biofuels in the Netherlands", *Technological Forecasting and Social Change*, vol. 76, no. 8, pp. 1003–1020, 2009.

[TAN 15] TANGUY C., MARTIN M., "Le développement durable dans la filière viticole bourguignonne: stratégies et pratiques", *Innovations – Revue d'Economie et de Management de l'Innovation*, vol. 46, pp. 141–160, 2015.

[TEM 11] TEMRI L., "Innovations technologiques environnementales dans les petites entreprises: proposition d'un modèle d'analyse", *Innovations. Cahiers d'Économie de l'Innovation*, vol. 34, pp. 11–36, 2011.

[TRO 09] TROTIGNON R., *Comprendre le réchauffement climatique*, Pearson Education, 2009.

[TRU 02] TRUFFER B., METZNER A., HOOGMA R., "The coupling of viewing and doing: strategic niche management and the electrification of individual transport", *Greener Management International*, vol. 37, pp. 111–124, 2002.

[TRU 12] TRUFFER B., MARKARD J., BINZ C. *et al.*, "Energy innovation systems. Structure of an emerging scholarly field and its future research directions", EIS Radar Paper, November 2012.

[TSA 12] TSANG F., PEDERSEN J.S., WOODING S. *et al.*, "Bringing the electric vehicle to the mass market: a review of barriers, facilitators and policy interventions", the RAND Europe working paper series, 2012.

[UDA 08] UDAYASANKAR K., "Corporate social responsibility and firm size", *Journal of Business Ethics*, vol. 83, pp. 167–175, 2008.

[UNR 00] UNRUH G.C., "Understanding carbon lock-in", *Energy Policy*, vol. 28, pp. 817–830, 2000.

[URR 05] URRY J., *Sociologie des mobilités*, Armand Colin, p. 252, 2005.

[VAN 87] VAN DEN BELT H., RIP A., "The Nelson–Winter–Dosi model and synthetic dye chemistry", BIJKER I.W., HUGHES T.P., PINCH T. (eds), *The Social Construction of Technological Systems*, MIT Press, Cambridge, pp. 135–158, 1987.

[VAN 93] VAN DE VEN A.H., "The development of an infrastructure for entrepreneurship", *Journal of Business Venturing*, vol. 8, pp. 211–230, 1993.

[VAN 98] VAN LENTE H., RIP A., "The rise of membrane technology: from rhetorics to social reality", *Social Studies of Science*, vol. 28, no. 2, pp. 221–254, April 1998.

[VAN 99] VAN DEN ENDE J., KEMP R., "Technological transformations in history: how the computer regime grew out of existing computing regimes", *Research Policy*, vol. 28, no. 8, pp. 833–851, 1999.

[VAN 07] VAN DEN HOED R., "Sources of radical technological innovation: the emergence of fuel cell technology in the automotive industry", *Journal of Cleaner Production*, vol. 15, pp. 1014–1021, 2007.

[VAN 10a] VAN ALPHEN K., HEKKERT M.P., TURKENBURG W.C., "Accelerating the deployment of carbon capture and storage technologies by strengthening the innovation system", *International Journal of Greenhouse Gas Control*, vol. 4, no. 2, pp. 396–409, 2010.

[VAN 10b] VAN BREE B., VERBONG G.P.J., KRAMER G.J., "A multi-level perspective on the introduction of hydrogen and battery-electric vehicles", *Technological Forecasting and Social Change*, vol. 77, pp. 529–540, 2010.

[VEN 05] VENTURI M., BALZANI V., GANDOLFI M.T., "Fuels from solar energy. A dream of Giacomo Ciamician, the father of photochemistry", *Proceedings of the Solar World Congress*, Orlando, Florida, 2005.

[VER 07] VERBONG G., GEELS F., "The ongoing energy transition: lessons from a socio-technical, multi-level analysis of the Dutch electricity system (1960–2004)", *Energy Policy*, vol. 35, no. 2, pp. 1025–1037, 2007.

[VER 12] VERBONG G., LOORBACH D. (eds), *Governing the Energy Transition: Reality, Illusion or Necessity?*, Routledge, 2012.

[VIT 12] VITALI G., "Agreements and joint ventures in the electric vehicle industry", in CALABRESE G. (ed.), *The Greening of the Automotive Industry*, Palgrave MacMillan, pp. 225–239, 2012.

[VOS 06] VOSS J.-P., BAUKNECHT D., KEMP R., *Reflexive Governance for Sustainable Development*, Edward Elgar, Cheltenham, 2006.

[VOS 09] VOSS J.-P., SMITH A., GRIN J., "Designing long-term policy: rethinking transition management", *Policy Sciences*, vol. 42, pp. 275–302, 2009.

[WES 14] WESSELING J.H., FABER J., HEKKERT M.P., "How competitive forces sustain electric vehicle development", *Technological Forecasting and Social Change*, vol. 81, pp. 154–164, 2014.

[WIL 01] WILSON C., TISDELL C., "Why farmers continue to use pesticides despite environmental, health and sustainability costs", *Ecological Economics*, vol. 39, no. 3, pp. 449–462, 2001.

[WIN 02] WINSKEL M., "When systems are overthrown 'The Dash for Gas' in the British electricity supply industry", *Social Studies of Science*, vol. 32, no. 4, pp. 563–598, 2002.

[WIR 11] WIRTH S., MARKARD J., "Context matters: how existing sectors and competing technologies affect the prospects of the Swiss Bio-SNG innovation system", *Technological Forecasting and Social Change*, vol. 78, no. 4, pp. 635–649, 2011.

[WOE 15] WOESNER R. (ed.), *Cop 21 – Déprogrammer l'apocalypse*, Atlande, 2015.

Index

S, T, W

Other titles from

iSTE

in

Innovation, Entrepreneurship and Management

2016

BARBAROUX Pierre, ATTOUR Amel, SCHENK Eric
Knowledge Management and Innovation (Smart Innovation Set – Volume 6)

BEN BOUHENI Faten, AMMI Chantal, LEVY Aldo
Banking Governance, Performance And Risk-Taking: Conventional Banks Vs Islamic Banks

BOUTILLIER Sophie, CARRÉ Denis, LEVRATTO Nadine
Entrepreneurial Ecosystems (Smart Innovation Set – Volume 2)

GALLAUD Delphine, LAPERCHE Blandine
Circular Economy, Industrial Ecology and Short Supply Chains (Smart Innovation Set – Volume 4)

GUERRIER Claudine
Security and Privacy in the Digital Era (Innovation and Technology Set – Volume 1)

MEGHOUAR HICHAM
Corporate Takeover Targets

MONINO Jean-Louis, SEDKAOUI Soraya
Big Data, Open Data and Data Development (Smart Innovation Set – Volume 3)

MOREL Laure, LE ROUX Serge
Fab Labs: Innovative User
(Smart Innovation Set – Volume 5)

2015

CASADELLA Vanessa, LIU Zeting, DIMITRI Uzunidis
Innovation Capabilities and Economic Development in Open Economies
(Smart Innovation Set – Volume 1)

CORSI Patrick, MORIN Dominique
Sequencing Apple's DNA

CORSI Patrick, NEAU Erwan
Innovation Capability Maturity Model

FAIVRE-TAVIGNOT Bénédicte
Social Business and Base of the Pyramid

GODÉ Cécile
Team Coordination in Extreme Environments

MAILLARD Pierre
Competitive Quality and Innovation

MASSOTTE Pierre, CORSI Patrick
Operationalizing Sustainability

MASSOTTE Pierre, CORSI Patrick
Sustainability Calling

2014

DUBÉ Jean, LEGROS Diègo
Spatial Econometrics Using Microdata

LESCA Humbert, LESCA Nicolas
Strategic Decisions and Weak Signals

2013

HABART-CORLOSQUET Marine, JANSSEN Jacques, MANCA Raimondo
VaR Methodology for Non-Gaussian Finance

2012

DAL PONT Jean-Pierre
Process Engineering and Industrial Management

MAILLARD Pierre
Competitive Quality Strategies

POMEROL Jean-Charles
Decision-Making and Action

SZYLAR Christian
UCITS Handbook

2011

LESCA Nicolas
Environmental Scanning and Sustainable Development

LESCA Nicolas, LESCA Humbert
Weak Signals for Strategic Intelligence: Anticipation Tool for Managers

MERCIER-LAURENT Eunika
Innovation Ecosystems

2010

SZYLAR Christian
Risk Management under UCITS III/IV

2009

COHEN Corine
Business Intelligence

ZANINETTI Jean-Marc
Sustainable Development in the USA

2008

CORSI Patrick, DULIEU Mike
The Marketing of Technology Intensive Products and Services

DZEVER Sam, JAUSSAUD Jacques, ANDREOSSO Bernadette
Evolving Corporate Structures and Cultures in Asia / Impact of Globalization

2007

AMMI Chantal
Global Consumer Behavior

2006

BOUGHZALA Imed, ERMINE Jean-Louis
Trends in Enterprise Knowledge Management

CORSI Patrick *et al.*
Innovation Engineering: the Power of Intangible Networks

Printed and bound by CPI Group (UK) Ltd, Croydon, CR0 4YY

13/09/2026

14961020-0008